Citizens, Cops, and Power

Citizens, Cops, and Power

Recognizing the Limits of Community

STEVE HERBERT

The University of Chicago Press   Chicago and London

STEVE HERBERT is associate professor in the Department of
Geography and in the Law, Societies, and Justice Program at the
University of Washington. He is the author of *Policing Space:
Territoriality and the Los Angeles Police Department* (1997).

The University of Chicago Press, Chicago 60637
The University of Chicago Press, Ltd., London
© 2006 by The University of Chicago
All rights reserved. Published 2006
Printed in the United States of America

15 14 13 12 11 10 09 08 07 06   1 2 3 4 5

ISBN: 0-226-32730-2 (cloth)
ISBN: 0-226-32731-0 (paper)

Library of Congress Cataloging-in-Publication Data

Herbert, Steven Kelly, 1959–
  Citizens, cops, and power : recognizing the limits of community /
Steve Herbert.
    p. cm.
  Includes bibliographical references and index.
  ISBN 0-226-32730-2 (cloth : alk. paper)—ISBN 0-226-32731-0 (pbk. :
alk. paper)
    1. Community organization—United States. 2. Community life—
United States. 3. Community policing—United States. 4. Police-
community relations—United States. 5. Crime prevention—Citizen
participation—United States. I. Title.
  HN90.C6H47 2006
  363.2'3'0973—dc22
                                        2005021525

♾ The paper used in this publication meets the minimum requirements
of the American National Standard for Information Sciences—
Permanence of Paper for Printed Library Materials, ANSI Z39.48-1992.

To the memory of

F. HARVEY HERBERT

Contents

Acknowledgments

It is fitting to begin a book about community by emphasizing the communal support upon which I so heavily relied. The list of friends and colleagues who assisted me along the way is longer than I can note here, but their collective support made the process more enjoyable and greatly improved the product.

The idea for this project emerged while I was on the faculty at Indiana University. There I enjoyed the support of several colleagues, most notably Kip Schlegel, Ed McGarrell, Steve Chermak, and Ellen Dwyer. Kip and Ed were particularly encouraging about this project in its very early gestation. Doris Marie Provine provided constructive suggestions during the process of soliciting resources from the National Science Foundation, which provided grant funds through its Law and Social Science Division.

The best gift from that grant was the opportunity to hire Elizabeth Whitford as my research assistant. The caliber of the analysis that follows is deeply tied to the skill with which she conducted interviews with the citizens of West Seattle. Her many, and unteachable, conversational skills spurred the interviewees to speak openly and thoughtfully. Elizabeth was also a terrific sounding board and collaborative problem-solver. I cannot imagine a more fruitful relationship with a research assistant.

Those citizens who consented to interviews also deserve my warm thanks for agreeing to talk to us about their communities. Their openness was both refreshing and illuminating. Similarly open were the many officers of the Seattle Police Department who engaged with this project. They patiently endured myriad questions about their practices, and engaged in ongoing conversations about police-community relations. My thanks to Chief Gil Kerlikowske for enabling the research to occur.

Colleagues at the University of Washington helped immensely with the tasks of analysis and writing. My fellow faculty members in the Law, Societies and Justice Program read some of these chapters in their early form; their constructive commentary led to sharper formulations and deeper analysis. It is a characteristic trait of these colleagues—Katherine Beckett, Rachel Cichowski, Angelina Godoy, Alexes Harris, George Lovell, Michael McCann, Jamie Mayerfeld, and Arzoo Osanloo—to be supportive and helpful. This is why LSJ is such a compelling intellectual home. Michael McCann deserves special thanks for bringing the LSJ enterprise into being and for providing additional resources for research assistance through the Comparative Law and Society Studies Center.

I am further blessed at UW to be part of the geography department, another vibrant group. Several colleagues there—Michael Brown, Mark Ellis, Lucy Jarosz, Vicky Lawson, Katharyne Mitchell, and Matt Sparke— bolstered my spirits in various ways during the writing process, and I thank them warmly. The geography department is also home to a group of terrific and energetic graduate students, some of whom I have the pleasure of working with closely. For keeping me engaged and on my toes, I thank Victoria Babbit, Elizabeth Brown, Jean Carmalt, John Carr, and Kris Erickson. Students in my graduate seminar read an earlier version of this manuscript and provided constructive commentary; for this, I thank John Carr, Dominic Corva, Caroline Faria, Keith Goyden, Serin Houston, Iza Hussin, Arda Ibikoglu, and Tuna Kuyucu.

Assistance came from beyond the UW campus as well. Several folks— Nick Blomley, Bernard Harcourt, Peter Manning, Eugene Paoline, Jonathon Simon, and an anonymous reviewer—read the entire manuscript and provided many helpful suggestions. Others provided ongoing support, including advice on book titles and other pertinent issues; for these gifts, I thank Dydia DeLyser, Blake Gumprecht, Dave McBride, Carrie Mullen, Garth Myers, Michael Musheno, and Joe Nevins.

The best gifts of all came from the community that is my immediate family—Katherine Beckett and Jesse and AnnaRose Beckett-Herbert. Their love, laughter, and energy sustain me and keep me whole. I am honored each day to come home to them.

Sadly, I can no longer come home to see my father, Harvey Herbert, who died during the fieldwork for this project. As a model of quiet intelligence and concerned community action, he was unparalleled. It is a privilege to say that I am his son, and to be able to dedicate this book to his memory.

Introduction

On a weekday evening, I arrive a few minutes late to a meeting in a church basement in a diverse, crime-ridden neighborhood in Southwest Seattle. As I enter, a lieutenant for the Seattle Police Department is reviewing a document his office created recently as part of an application for resources from the federal government's Weed and Seed Program. If successful, the application will bring federal monies for both "weeding" and "seeding." Resources would be allocated first for intensive law enforcement to arrest and incarcerate known criminals, and second to meet such basic needs as housing, jobs, and education. The theory is that a neighborhood, like a garden, can be restored and replenished, and that serious crime issues can thereby abate.

The crowd around the table is not large. Counting the lieutenant, there are six local government officials. Seven others work with nonprofit organizations that stand poised to assist in the "seeding." There are just six citizens.

The lieutenant makes numerous points during his presentation, but two in particular are emphasized. One is that residents should not develop excessively high expectations of the Weed and Seed operation. In particular, they should expect no more than a 10 percent reduction in crime. He acknowledges that an earlier version of the document suggested that a 30 percent reduction in crime was possible, but he dismisses that figure as unrealistic. If the operation were that effective, he says, he would be eligible for a "cabinet-level position." He says further that even a 10 percent reduction is ambitious. The second point he stresses is his personal sense of responsibility for the project. He describes it as "my baby" and tells the audience that they "are in good hands."

After his presentation, the discussion opens. Many express a hope that the police's operations not be heavy-handed. For example, a representative of a youth service organization wants the police to use "weed" money to fund an officer to work strictly with youth. He argues that this would place emphasis where it belongs—on youth—and would invite something other than a strict law-enforcement strategy. His experience leads him to conclude that youth officers are creative in their approach: they find diverse means to help steer youth clear of criminal activities, and they involve themselves deeply in the familial and community contexts out of which problems spring. Another citizen asks that the police avoid strong-arm tactics. Such tactics, he argues, do little to build trust between the community and the police. Yet another hopes that officers will spend time simply hanging out in the community, getting to know residents better and building greater trust. Along these lines, one resident suggests that officers patrol not in cars, but on foot or on bicycles. A more impassioned citizen asks: "Do we have clout to do something other than just throwing people away? Are we going to make it positive for kids, or are we just having more meetings?"

The lieutenant acknowledges but does not directly engage this line of discussion. He does say that paying an officer to focus strictly on youth would "strip mine" his budget, and thus is unfeasible. Other than that specific comment, his responses consist of reassurances. He says that he will take ownership of the Weed and Seed operation and will hold himself accountable by attending all public meetings. Even though some of them do not know him, he argues, they should take comfort in his competence and willingness to engage.

Another focus of the discussion concerns the involvement of citizens in the ongoing oversight of the Weed and Seed operation. The convener of the meeting works for the local nonprofit crime prevention organization that will help oversee Weed and Seed. She mentions the need for various community groups, such as churches, to become involved, and she expresses a concern that the application document makes the process seem "very professional, and very top-down." She hopes that she and the other professionals involved will be seen as part of the group; she wants the people involved in Weed and Seed to be responsive to the desires of the citizenry, not vice versa. "Let's be totally different," she says.

At no point, however, does anyone propose any means to make "being different" possible. Indeed, what is most notable is the already low number of citizens in attendance. Of these six, only three actually live in the target area. And all six are members of a small group who attend most such meetings. Because I have observed many of those meetings in the months preceding this one, I am on a first-name basis with all of these "regulars." I talk with some of them after the meeting, and their sense of trepidation is palpable. For months, they have been promised that the Weed and Seed operation will address their ongoing concerns about crime and other symbols of the area's distress. Now, they are not so sure. Where, they continue to wonder, is their community headed?

The God Word

As Bell and Newby note, the word "community" is indeed a God word in American discourse.[1] The ideal of reasonably cohesive local units capable of self-government has a long history in the United States, promoted by such seminal thinkers as Thomas Jefferson and noted by such early commentators as Alexis de Tocqueville. The hope is for residents in particular localities to gather as a political entity, to know together the problems they must confront, and to decide how to ameliorate them. In this way, citizens learn more about themselves and each other and experience the unique sense of political effectiveness that comes through communal action. Through this political work, isolated individuals can create an efficacious collective and can control themselves with minimal outside help. In Michael Sandel's words: "When politics goes well, we can know a good in common that we cannot know alone."[2]

The notion of community is often central to these visions of decentralized politics, and its salience continues into the present day. In fact, it has received a considerable boost in recent years through a movement loosely defined as communitarianism. One thrust of this movement is in political theory, where thinkers like Michael Sandel, Alistair McIntyre, Charles Taylor, and others seek to challenge the hegemonic position of liberalism, which they believe overemphasizes the rights-bearing individual. Another thrust of this is more overtly political, with public intellectuals like Amitai Etzioni and Don Eberly arguing for the need to restore the moral fabric in American society through the reinforcement of communally held values.

The criminal justice system is one of the many places where one sees evidence of this trend, most notably in policing. Something called "community policing" is now the stated practice of police departments across the United States and throughout much of the world. What community policing means in practice is often unclear; different police departments do different things.[3] Despite this variation, community policing programs generally seek to help police departments to improve their connections with citizen groups and to decentralize their operations to make better use of input from those groups. This might mean increasing foot or bicycle patrols to make officers more accessible. It might mean ensuring that officers work the same beat for an extended period, so they can establish ties to residents. And it might mean pushing decision-making down the hierarchy, so that projects for local communities can be developed by officers closest to the action rather than by less-informed superiors in a distant, downtown office.

The language of "partnerships" is frequently invoked here, to capture the sense in which community groups and police officials are to come together to isolate and solve problems collaboratively. The logic is sensible. Any serious effort to combat crime must include both informal and formal social control. And, in coming together to fight crime, communities can know the ideals of political efficacy and self-determination long treasured in American culture.

Weed and Seed ostensibly fits this bill. It seeks to address significant issues in distressed communities, namely crime and the dearth of basic social services, through a partnership between government and citizen groups. It mandates levels of citizen involvement, from resident involvement on oversight committees to required public meetings such as the one just described. The premise of the program, on its face, is unobjectionable: that such partnerships can best stanch disorder in distressed neighborhoods and create new forms of order. Because both formal and informal sectors need to play significant roles in addressing phenomena like crime that reduce a neighborhood's quality of life, they must be brought together at the start.

But what does it mean to involve something called the "community" in this way? What does this community look like? What is it, actually? Can distressed neighborhoods be accurately described as communities? If so, upon what basis? Can they organize in any cohesive, representative, or effective fashion? Who can claim the right to speak for a community? And even if a community can be legitimately named, effectively organized, and accurately represented, just how can it interact with state agencies such as the police? What is the ideal division of labor between communities and the police, between the informal and the formal? Beyond these ideals, are the police, as a political agency and a cultural creation, able to hear what the community has to say? What forms of citizen input are credible to the police? Do the police's professional norms and cultural practices keep citizens at arm's length, or can something approaching a genuine partnership between cops and communities be created?

The Challenge of Community

As my story above illustrates, these are not merely rhetorical questions. Two aspects of the meeting were particularly striking and were replicated in numerous other such meetings. One is that the extent of citizen involvement in this particular "partnership" was hardly robust, and consisted of a set of regulars who are demographically quite notable: unlike most of their neighbors, they are white, middle-aged property owners. The second was the unwillingness of the lieutenant to engage directly suggestions from those few citizens in attendance. While not confrontational, the lieutenant nevertheless deflected or ignored recommendations about how he should orient the weeding operations. As an alleged partnership, this interaction left much to be desired; in the name of community, nothing too impressive is transpiring. Local groups display little evidence of diverse participation or political efficacy, and state actors seem deaf to what input they do receive.[4]

Despite the understandable desire for communal self-determination, the assertion of the necessity of "community" involvement in efforts to address such problems as crime is not a straightforward one. Perhaps, as Bell and Newby suggest, its invocation is meant to entice us to abase ourselves before it, and thereby to legitimate what is done in its name.[5] But the ideal and actual role of community in local governance deserves considered and critical attention. I seek to provide that attention in the pages that follow.

I focus on two questions: (1) Can and should something called "community" be treated as a legitimate, important, and effective political actor, as it could potentially be in projects constructed in the name of community policing? And (2) if it can, how can or should "community" interact with state agencies such as the police?

There are numerous reasons for skepticism about invocations of community in efforts to address local problems. One set of these concerns comes from political theory, largely that body of political theory that defends the political ideology of liberalism. Here, there is a fear that the invocation of "community" can sanctify majoritarianism, which could limit the rights of individuals or minority groups. Better, liberals assert, to preserve those rights to ensure that political pluralism is possible. Otherwise, an alleged communal normative system could suppress alternate constructions of the good life. Though they start from a different position, poststructuralist and feminist theorists arrive at the same conclusion: that community is necessarily built on oppositions—between an "us" and a "them"—and thus smuggles in oppression in the name of localized democracy. Even if community can be invoked for arguably beneficent causes, it can also work as an act of closure, and thus must be held in perpetual suspicion.

Another set of concerns is more sociological, focused on the realities of contemporary urban communities. Perhaps urban residents do not experience a sense of community in their neighborhoods; their social network may not be localized in a particular place, and the demands of daily life and the realities of pluralism may render communal activity unlikely. As a consequence, neighborhood groups struggle to coalesce—people are either otherwise engaged or unable to cross the demographic barriers that divide them. And even if these barriers could be hurdled, the cultural valorization of individualism arguably inhibits the drive to organize collectively.

Further, even if community governance could be justified and made possible, what then happens when citizen groups face a state agency such as the police? Here, there are also significant challenges, emerging again from political theory and empirical analysis. In terms of theory, there is considerable debate about just how community groups, and society more broadly, should stand in relation to the state. Should state agencies like the police be controlled by the citizenry, as the general thrust of democracy suggests? Or should state agencies stand somehow separate from society, as in the ideals of liberalism, ruled more dispassionately and neutrally by formalized laws, the better to ensure the sanctity of individual rights? Additionally, might not the police's emphasis on their professional norms, and their attendant desire to assert their unique role in fighting crime, work to reinforce rather than mitigate a sense of separation? Or perhaps the state should be best understood as principally generative of society and its communal groups, capable of determining the welfare of communities through its policies and principally defining what, in political terms, community actually is? In short, in terms of its stance toward community, should the state be understood as *subservient, separate,* or *generative*? As we will see, each of these understandings of state-society relations has some legitimacy, and this makes it difficult to outline an ideal relationship between the citizenry and the police.

But these theoretical questions are not all. We must also understand the social and political world of the police, the better to appreciate how "community" is recognized by officers. How do police officers comprehend the information they receive from urban residents? And what underlies the responses they are likely to make? In short, to understand more completely whether there is viability for "community" in urban political projects that involve crime and related quality of life issues, one must undertake a sociological analysis of the police.

Community in Practice

The challenge of assessing the political viability of community in urban projects like crime reduction is thus a matter of both theory and empirical analysis. Political theory provides normative arguments that outline both the ideal role of communal groups and their preferred relationship with the state, sociological theory with an account of those factors of urban life that might promote or suppress the drive to organize. A more concrete analysis enables us to understand whether any of these normative ideals possesses any traction on the ground and to understand what social factors promote or thwart communal action.

In what follows, I move on each of these fronts via a theoretically informed qualitative case study of a set of neighborhoods in West Seattle. Set on a peninsula and only accessible to the rest of the city via three bridges, West Seattle is a distinct area within the metropolis, with a strong sense of identity. Many residents describe it as an intimate, familiar place. As one resident, Dee,[6] described it, "West Seattle in general feels like a small community, and so you almost feel like you're in your own small town." Another resident, Sally, noted that West Seattle possesses a "more secluded feel, more of a neighborhood feel" than other areas in the city. Others referenced their ability to know shop owners on a first-name basis or the chance to encounter friends at the weekly farmers' market. This latter, according to Stephanie, another West Seattlite, is "a place where hundreds of people go and gather together every week. It's a real community center, all kinds of different people from the neighborhood come." Given its isolation, identity, and intimacy, West Seattle is a place where a sense of community might emerge with greater force than in many urban locales.[7]

Yet there is significant variance. Marked variations in West Seattle's topography produce corresponding variations in real estate pricing. High-valued property is found along the beach or on high bluffs with scenic views.[8] Lower values are found in the valleys and in the flatlands to the south. To capture this variation, I chose three contiguous yet diverse police beats. One of these, which I call "Beachland," lies along the western edge of West Seattle. It includes homes with median incomes well above the Seattle average. The vast majority of these homes are owner-occupied. The population is also about 90 percent white. Not surprisingly, its rates of reported crime are low, as are the number of calls for police service.[9]

The second area, "Midlands," is significantly more diverse. Located in the center of the peninsula, this area possesses many upper-middle-income residents living in owner-occupied homes in areas that are predominantly white. But it also includes a large public-housing project, "Blufftop," whose residents are poor and ethnically diverse. The facility, composed of small, detached units, sprawls across several acres stitched together by a meandering street pattern. It generates a steady flow of calls for police service, in part because of a long history of outdoor drug sales on the grounds.[10] The overall crime pattern for Midlands, as measured by arrests, places it well above the average for Seattle. It has particularly high numbers of reports of burglary, auto theft, domestic assault, and narcotics. Police officers attribute these high numbers in Midlands to the presence of Blufftop.

The third area, "Flatlands," also has a high crime rate, and it generates the highest number of calls for service of any SPD beat. It has high numbers of reports for arson, auto theft, burglaries, domestic assaults, homicide, and narcotics. The demographic pattern is mixed, with two largely middle-class neighborhoods—which I call "Westside" and "Eastside"—bracketing a more diverse and poorer neighborhood, "Centralia." The former two possess significantly more owner-occupied homes, and are majority white. By contrast, Centralia's homes are 45 percent owner-occupied, and the white population only barely exceeds 50 percent. African Americans, Asians, and Latinos each account for about 15 percent of the population, with American Indians providing the bulk of the rest.

This demographic variance within a tight-knit urban locale offers an unusual opportunity to unearth the range of factors that might either prohibit or promote both effective communal governance and productive community-police interactions. I explore these dynamics through the use of three sets of qualitative data. The first comes from extensive, semistructured interviews with forty-six residents in the three police beats, conducted primarily by a trained research assistant. These interviews enabled residents to discuss "community" at some length: what the term meant to them; whether their neighborhood warranted the label; what they hoped from community; what they sought to contribute to it; what role it could play in neighborhood betterment. The interviews also compelled residents to describe what problems, if any, their neighborhood faced, and the role they envisioned for the state in resolving those problems. In this vein, they were asked specifically to discuss problems of crime and disorder and to articulate their vision of the role of police. Residents also assessed the police's actual practices. The interviews covered a schedule of questions, but they were conducted in a loose and improvised fashion, to give residents the conversational freedom to expand upon particular issues. The interviews typically lasted more than an hour and generated transcripts that averaged thirteen single-spaced pages.

The population of interviewees was diverse. Half had lived in their homes for seven years or longer, half for less. The largest group, 18, were in their forties; 14 were younger (including 7 in their teens), and 14 were older. Forty-two percent earned less than $35,000 a year, 26 percent earned between $35,000 and $65,000, and 14 percent earned more than $65,000. (Eighteen percent declined to provide income information.) Twenty-six of the interviewees (56 percent) were white; 10 (22 percent) were African-American or African; 3 (6.5 percent) were Latino; 3 (6.5 percent) were Asian ; and 4 (9 percent) were American Indian.

There was over-sampling in two categories. All but seven of the interviewees came from either Midlands or Flatlands. Because its crime pattern was minimal, Beachlands was a less instructive place to discern patterns of communal governance and of police-citizen interaction. Indeed, none of the Beachland residents had any serious concerns about crime, and all of them reported minimal contact with the police. By contrast, many Midlands and Flatlands residents reported considerable concern about crime and regular interaction with police officers. These neighborhoods provided greater opportunities to assess the promise of community policing.

Anti-crime activists were the second over-sampled group. About one-third of the interviewees described a high level of past or present involvement in collective attempts to reduce crime. Such involvement positioned these respondents to assess the challenges and potential for productive communal efforts to work with the police to reduce crime.

The second set of data consists of field notes gathered during ride-alongs and interviews with members of the Seattle Police Department. The twenty-three ride-alongs were with three groups of officers with responsibility for one of the three beats. Thirteen were with patrol officers, who are "first responders" to calls for service and who regularly cruise the streets. Five were with sergeants, who supervise patrol officers, often in the field. And five were with members of the "Community Police Team." These officers are the principal public face of community policing in Seattle. Their main tasks are (1) to attend public meetings to provide information and solicit input; and (2) to monitor areas of ongoing concern, such as a home or street corner suspected of hosting drug trafficking. The ride-alongs were an opportunity to learn how officers conceive of "community" and to witness how those conceptions shaped officers' approaches to the citizenry. Besides these ride-alongs, I conducted interviews with three lieutenants and two captains. The lieutenants were each responsible for some portion of the community police operation, and the captains headed the precinct that included the study police beats.[11]

The third set of data was collected at twenty-nine meetings that brought together police representatives and citizens. Most of these were regularly scheduled gatherings of neighborhood organizations or anti-crime groups, although some—such as the Weed and Seed meeting described above—were one-time, special meetings. These meetings offered a chance to learn what "community" means in practice, how it is that police and other government agents do—and do not—work together to confront problems of crime and disorder.

Together, these data enable an opportunity to assess the political weight community can support in projects like community policing. Residents assess community and its potential for political action; police officers describe how they like best to articulate with citizens; public gatherings show "partnerships" in action. Brought together in the same analytic project, these data can make plain the possibilities and pitfalls of seeking communal political regeneration through a project like community policing.

On Qualitative Case Study Research

It is important to stress that my goal is not to evaluate community policing for its "success," which is usually measured in terms of its effects on crime and fear of crime.[12] I seek instead to outline the normative assumptions that legitimate community policing and to assess these against the assumptions that residents and police officers employ in understanding themselves and each other. My interest, ultimately, is in the capacity for community policing to enable neighborhoods to generate localized democratic action and to garner the attention of the state.

This explains my use of qualitative data. Research that focuses on what "works" inevitably takes a strong quantitative tack, because it seeks to measure the impacts of police practice on some outcome, such as crime or fear of crime. My decision to foreground normative questions and to assess their tractability on the ground leads me down a different methodological path. The qualitative data I marshal here are the only appropriate choice for the questions I address. Only these data can enable an exploration of what community means to those who are to create and enact it, and of the processes through which it is, or is not, enacted.

One potential downside here, of course, is that I am confined to a single case study at a single point in time. One might then question my capacity to generalize the lessons of my analysis to other locales. Two points lessen considerably the bite of this line of critique. One is that the general finding I describe here—of minimal productive "partnerships" between cops and citizens—is not uncommon in other locales; indeed, it appears to be the norm.[13] Given that urban police departments in the United States face similar challenges and develop similar organizational profiles, there is every reason to believe that the widespread failure to create stronger partnerships derives from similar causes. What I offer here is an explanation for why this reality is so commonplace across locales. Other places may well differ in some respects from Seattle, but there are strong reasons to believe that the impediments I outline recur elsewhere.[14]

Secondly, West Seattle is a place where community policing might be expected to take root. If it cannot succeed there, then perhaps there are inherent problems with community policing that warrant the considered attention I provide. Seattle is therefore an ideal test case.[15]

One particular advantage of qualitative data is that it enables social meanings and processes to be rendered understandable through the actual words of residents and the actual practices of important actors like the police.[16] In the analysis that follows, I make considerable use of such data to illustrate my arguments as vividly as possible. To use data in this way, however, raises a legitimate concern about the representativeness of any particular quotation or incident. Although I did no statistical analysis to address directly the representativeness of resident comments or police behavior, I only use data to illustrate perspectives or patterns that were evident across multiple interviews or observations.

An additional issue to address at the outset is the use of the term "community" in this analysis, a point to which I now turn.

The Meaning of Community

Community policing, when put into practice, makes a conflation between community and neighborhood. There is an implicit presumption that urban neighbors share common problems of crime and disorder. They should therefore organize at the scale of the neighborhood and address their problems through productive relations with the police. Of course, police departments do recognize local groups that organize at larger scales, but the neighborhood scale is critical.

This conflation between community and neighborhood can be considered problematic. That is because many people do not understand community as spatially bounded; urban residents often seek community outside their neighborhood.[17] No longer necessarily tied to locales for critical supports such as employment, shopping, and child care, urbanites establish wide networks of social association, particularly if they are economically comfortable.[18] In terms of politics, even "grass roots"–oriented political organizers in American cities often work consciously to extend their lines of affiliation across manifold neighborhoods.[19] Indeed, to recognize that urban residents, as citizens, possess political agency is to underscore their capacity to choose the communities with which they affiliate.[20]

Yet this argument can be overstated. Territorial connections remain regnant for many urban dwellers.[21] Those who live next to one another, as Albert Hunter evocatively notes, are bound by "the common fate of shared space."[22] As Robert Sampson puts it: "Local community remains essential as a site for the realization of common values in support of social goods, including public safety, norms of civility and mutual trust, efficacious voluntary associations, and collective socialization of the young."[23] Most importantly for my analysis, the conflation between community and neighborhood is quite common in urban political projects.[24] If a group is organized around a localized problem such as crime, community is often treated as an analogue to neighborhood. This obviously occurs in the case of community policing, and in other instances where localized democracy is heralded. It is therefore essential to consider the extent to which a neighborhood is experienced as a community, and to ask whether any such community can bear the weight that projects like community policing place upon it. So, even if many urban residents do not necessarily circumscribe "community" territorially, many government agencies do, with the range of considerable consequences I explore.

Outline of the Book

My analysis suggests that community is all too often *unbearably light*, in two respects. First, community is not a sufficient support for the political responsibilities it is often meant to assume. Urban residents do not typically wish for their neighborhoods to act in a politically robust manner. Further, even when they are inclined to organize collectively, they see clear limits on the potential effectiveness of any neighborhood group. Thus, urban neighborhoods cannot bear the weight of responsibility that programs like community policing are meant to place upon them.

Second, communities are usually light in terms of political voice. Even when members of neighborhood groups are able to articulate a concern, either individually or collectively, they can expect to encounter a largely unresponsive state apparatus. This reality is hardly a function solely of the organizing ability of urban neighborhoods. Rather, it results primarily from the internal operations and cultural orientations of state agencies like the police. As I will make plain, various internal dynamics limit the capacity of the police to engage in genuine partnerships with community organizations. Community requests thus resonate less as a shout than as a whisper.

Given this political lightness, I suggest that community should not typically be seen as an effective carrier of our hopes for localized democracy. I use the analysis of the data to substantiate this claim in the chapters that follow. I start, in chapter 1, by comparing normative visions of community with those articulated by residents. Here I document a significant disjuncture between the normative and the empirical. Residents evince a surprisingly hegemonic vision of community, one that is not well captured by normative theorists. This suggests a need to reexamine the sources of legitimacy for communal governance.

Chapter 2 explores the political capacity of community in greater depth. Here, the residents outline various factors they cite to explain their pervasive pessimism about any such political potential. Not surprisingly, the constraints on communal efficacy fall more heavily on poorer neighborhoods. This suggests that efforts to devolve power to local communities might reinforce existing class-based differences.

Chapter 3 shifts the focus from the community to the state—in particular, to the police. I outline there how conflicting demands for legitimacy place the police in an awkward position vis-à-vis the citizenry, as different ideals—subservience, separation, generativity—push the police to pursue different paths to make themselves politically palatable. I show how these modes are evident in everyday actions, and how they are perpetually in tension with one another.

My analysis shows that the narrative of separation possesses great resonance in the police's social world and underwrites a robust resistance to community policing. I explore this further in chapter 4, where I examine those cultural orientations and practices of the police that most potently shape their approach to the citizenry. Collectively, these dynamics lead officers to resist many of the dicta of community policing, most notably the ideal of communal efficacy.

Chapter 5 examines citizen assessments of the police and of police-community interaction. I review how citizens implicitly accept the reality that the police are simultaneously subservient, separate, and generative, and also how they are frequently frustrated when subservience is thwarted. These assessments illustrate how separation and generativity often diminish citizens' voice and reduce their capacity to influence officer action.

Finally, in chapter 6, I review and summarize the analysis, with the aim of making broader claims about what the data tell us about the viability of community policing as an important instance of the broader goal of making neighborhood-based organizations viable political actors. I suggest a need to rethink—and perhaps abandon—community policing as a reform movement. Its promise is unlikely ever to be realized. Further, community policing raises provocative questions about the roles of the state, community, and the relation between the government and the governed. These questions are inescapable and deeply normative. They are thus worthy of our continued attention.

It is a mistake to believe that community can bear the political weight that projects like community policing place upon it, particularly in urban neighborhoods in distress. Further, the term "community" can often generate confusion, particularly in terms of its implications for state-society relations. Any way forward needs to take cognizance of the realities of contemporary urban life and of the normative confusion that surrounds the invocation of community.

The Terrain of Community

So, there's just different levels of involvement, and I think a strong community would be everybody knows everybody, everybody says hello to everybody. It's kind of like a feel-good 1950ish type of situation, and you know, growing up, I knew everybody on my block. In fact, not just the street, the block. I knew everybody's kids. They knew exactly who everybody was. Everybody's family knew everybody else's family. And there was such an involvement. Everybody went to the same school. You saw the same people, the same parents at these functions. It was just really great. . . . But that was the 60's. This is different now.

ANDREW, MIDLANDS RESIDENT

I know that sounds kind of silly, but I really feel it's important for everybody to experience what a community is supposed to be about. And some people don't have any idea what it means to go say hi to your neighbor. When I was a kid everybody knew everybody, but now with busing and logistics and working, it's different than it used to be in the old days. We don't have that same sense anymore, and I don't know how we as a city can get there.

JEAN, EASTSIDE RESIDENT

How can cities "get there," in Jean's words? How can they attain a situation where residents of urban neighborhoods know and appreciate one another? Is it even possible for urban families to have the experience recalled by Andrew, where they know "everyone else's family" in the immediate area in which they live, where there is "such an involvement" with neighbors' lives? What impediments lie in the way of these ideals? And even if these impediments could be removed, should these ideals be pursued?

The philosophy of community policing, one popular instance of a drive toward greater communal governance, presumes that these ideals can and should be pursued. Its legitimacy rests upon the apparently unassailable assertion that urban citizens should come together to exert dominion over their neighborhoods. Security is often a collective good, and thus it should sensibly be pursued and reinforced collectively, via a process that marries the formal weight of the police with the informal dynamics in particular locales. Once organized, communities should articulate a common voice to which policy makers should respond.

Two normative assertions are being made here. One is that communities should construct themselves as political actors; the other is that the state should recognize them as such. One normative vision concerns the political status of community, the other the state-community relation.

A critical interrogation of the desirability and possibility of such communal governance projects as community policing must necessarily step back and assess these normative visions. To evaluate community policing's possibilities and legitimacy, we must hold these often implicit assumptions up to scrutiny. In this chapter, I begin this process by assessing the various significances ascribed to the term "community."

When we gain analytic distance from the rosy scenario of cohesive and politically capable communities, two sets of questions assert themselves. One set involves this normative vision of community. What informs this vision? What are its assumptions, proscriptions, weaknesses? Are there various such visions, and how do they contrast? In short, how do we understand and assess various depictions of the ideal community?

The second set of questions addresses how community is actually envisioned by residents of urban communities. Even if certain normative visions of community possess luster, do they jibe with the version of community that residents envision and enact in their daily behavior? How does the normative compare with the actual?

I pursue each of these sets of questions in the four sections of this chapter. In the first section, I seek to explain the resilience of community as a social and political ideal and to document how it came to be idealized in the now hegemonic project of community policing. Much of the seductiveness of community policing rests upon a largely unexamined romanticization of the small-scale democracy that neighborhoods can ostensibly create.

In the second section, I explain how normative visions of community and its political potential are by no means uniform. I do this by elaborating three principal normative visions. Two of these develop what I term a "thick" version of community—they each desire for community a robust role in social life—although they differ in an important way. The third is more a "thin" version, wary as it is of the exclusivity and majoritarianism community might entail. I review these three visions by explicating the assumptions upon which they build and the proscriptions they assert.

The third section draws extensively upon the interview data and reviews how the respondents talked about community. It becomes apparent that the common vision they express does not conform neatly with any of the above; it is a vision that is neither especially thick nor especially thin. What people want is a fundamental sense of connection and familiarity, but not in the pursuit of any larger set of values or political ideals. Rather, residents desire a feeling of security that derives from knowing one's neighbors well enough to be able to predict their actions. Their shared inhabitation of neighborhood space means that residents experience a collective need to protect each other from needless vulnerability, but this does not translate into a broader political project.

In the final section, I again mine the resident interviews to outline those factors that citizens believe maximize their feelings of security within the spaces they inhabit. Most critical here is property ownership. Residents hope for neighbors who demonstrate permanence and pride in ownership, qualities they believe renters commonly lack. Other glues include children, schools, and regular outdoor encounters. These characteristics, importantly, are more commonly found in neighborhoods of economic advantage, a point that will emerge with greater force in chapter 2, where I consider the hindrances to community cohesiveness and political effectiveness.

For now, the goal is to contrast the normative and the actual, the idealized visions of community and those articulated by citizens. As I demonstrate, the discrepancy between romanticized depictions of community governance and the actual desires of citizens is significant. This discrepancy should give us considerable pause in assessing projects like community policing, and should also cause us to wonder if urban neighborhoods can bear the weight of extensive self-governance. To see why this is the case, I begin by exploring just why community, and community policing, possess such political potency.

Why Community? Why Community Policing?

And it is easy to see why the word "community" feels good. Who would not wish to live among friendly and well-wishing people whom one could trust and on whose words and deeds one could rely? For us in particular—who happen to live in ruthless times, times of competition and one-upmanship, when people around us seem to keep their cards close to their chests and few people seem to be in any hurry to help us, when in reply to our cries for help we hear admonitions to help ourselves, when only the banks eager to mortgage our possessions are smiling and wish to say "yes," and even they only in their commercials and not their branch offices—the word "community" sounds sweet. What that word evokes is everything we miss and what we lack to be secure, confident and trusting.

ZYGMUNT BAUMAN, *COMMUNITY: SEEKING SAFETY IN AN INSECURE WORLD*

As Zygmunt Bauman suggests, community is clearly a word that feels good. It certainly sits at the center of many normative visions of the good life. It implies, at minimum, a degree of fundamental social connection that provides a shield against the unpredictable vagaries of modern existence. In more expansive visions, it provides a vehicle through which we can develop and actualize a range of goods and values: we can draw closer to God; we can develop our potentials as a musician, a writer, a parent; we can strive collectively to assist others. In terms of politics, community can serve as a principal means to protect an existing way of life against possible change. Alternatively, community can be a tool to try to create arrangements that more closely actualize such communally held values as justice, equality, and fairness.

Community is an elusive yet potent ideal in American social and political discourse. It is a focus of everyday concern, of sociological analysis, of efforts to revitalize politics. It is something potentially to be experienced daily, as one engages others in the minutiae of daily life and the shared responsibilities of raising children, maintaining property, and tending to the disadvantaged. It can reinforce connection as it helps people to meet basic needs or to pursue common interests. It can provide senses of belonging and direction. And it can be mobilized by a neighborhood to address such collective concerns as crime, property maintenance, and development. In its social dimension, it provides a communal grounding; in its political dimension, a collective heft for a group's interests.

Both of these ideals possess longstanding legitimacy. As the interviewees quoted above indicate, there is a frequently expressed nostalgia for the "good old days," when one knew one's neighbors via relations of familiarity and stability. This nostalgia is not just popularly expressed, but is a central theme in modern sociology. For early sociologists like Ferdinand Tönnies and Émile Durkheim, the challenge of modernity concerned the shift from close-knit rural societies to more socially differentiated urban centers, from *Gemeinschaft* to *Gesellschaft*, from mechanical to organic solidarity. As Nisbet noted, community is "the most fundamental and far-reaching of sociology's unit-ideas."[1] Much work in urban sociology has focused on the extent to which something called "community" can remain extant in the modern city, and if so, whether that community is localized or spatially dispersed.[2] Certainly, a sense of loss motivates many analyses undertaken in the name of communitarianism, a movement of both theory and politics that seeks to restore and make politically legitimate the bonds arguably only produced and reinforced via communal connections. From both academic and popular literatures, one hears frequent complaints about how the hegemonic status of individualism and the various complications of daily life increasingly leave urban residents disconnected from one another, and hence poorly able to act as a palliative or political collective.

Why this persistent longing for community? Part of the explanation must lie with our basic need for social connection. Community can thus be understood as an end in itself, a forum for making friends and establishing bonds. Further, such a community can help us meet basic needs. Who does not want a neighbor from whom one can cadge the occasional cup of sugar, the half hour of emergency babysitting, the advice or the tool to make a home improvement project successful? Beyond this, communal groups can be the indispensable fora for the realization of treasured values or the development of varying interests. In this way, communities are necessary for the ongoing pursuit of a life of meaning and exploration.

Political efficacy is one possible goal communities can pursue. Perhaps, some suggest, there is an ingrained—and too frequently undeveloped—desire for humans to experience a sense of self-governance. Just as humans are social creatures, they are implicitly political ones as well, with an innate need to exercise as much control as possible over their own lives. This line of argument is captured well in Benjamin Barber's insistence on what he terms "strong democracy."[3] Barber argues that citizens need to be able to govern themselves in at least some public matters at least some of the time. Without such political participation, he argues, "Women and men cannot become individuals."[4] Given the intrinsic interdependencies of human existence, this participation should occur through communal organizations. Discussions and debates within such organizations present opportunities for individual members to enlarge their understandings of one another and to grow and develop as they seek to forge ways forward together.[5] In such a fashion, citizens can develop together a politics of transformation, invention, and creation.[6] To increase the possibility for such communal participation, it makes considerable sense to devolve political authority to the lowest possible level. As political units decrease in size, their senses of solidarity and effectiveness should increase. Such smaller-scale institutions are more likely to represent the interests of the group, and thus to improve the sense of connectedness individuals feel to the collective.[7]

From this perspective, then, the good is best pursued through politics, namely through active citizenship in a virtuous, localized community. Jeffrey Berry summarizes this ideal well: "To the most degree feasible, we should engage in face-to-face democracy, working with our neighbors to govern ourselves rather than relying on elected representatives to make decisions on our behalf. Face-to-face participation will make us better citizens by educating us about our communities and teaching us to be tolerant and cooperative."[8] Such an ideal of collective efficacy is regnant across the political spectrum. Leftists, moderates, and conservatives all embrace local groups as a key to political regeneration, as central to any strategy of more overarching change.[9] For a range of political goods, in a range of places, small-scale collectives are seen as a viable if not indispensable means for ensuring that citizens are able to advance and protect their interests vis-à-vis larger social forces that might otherwise swamp them.

The Case for Community Policing

Security is one good that localized collectives might sensibly pursue. Problems of crime and disorder possess a geography, and thus residents of a given neighborhood will usually experience threats collectively.[10] Thus, they should sensibly seek protection from threats as a collective. Further, there is much evidence that informal social control efforts are critical contributors to a neighborhood's degree of security.[11] Such informal control can be expressed through various means–gossip, avoidance, ostracism, surveillance, scoldings—and can work to help maintain order, even in distressed neighborhoods.[12] An organized neighborhood can also exert pressure on the police and other components of the formal social control apparatus to help mitigate or eliminate an ongoing situation of crime or disorder. It is thus quite understandable that "community" makes frequent appearances in efforts to address the causes and consequences of crime.[13] Community policing is a popular and well-publicized example of such efforts.

Community policing's initial formulations grew out of concerns about poor police-community relations, particularly in urban neighborhoods of concentrated disadvantage.[14] It was arguably not by accident that many riots in America's cities in the 1960s were ignited by a tense police-citizen encounter. This tense dynamic was likely exacerbated by the police's adherence to the then hegemonic ideal of police organization, the professional model. As professionals, police understood themselves as the uniquely qualified experts in crime control who swooped onto crime scenes and deployed their tactics in a detached and efficient way.[15] Citizens were adjuncts to this process, brought in only under the precise guidelines laid down by the police: the "just the facts, ma'am" approach of that prototypical police professional, Sergeant Joe Friday of TV's *Dragnet*.

But this aloof, sometimes brusque style had two distinct disadvantages. First, it alienated the police from communities, particularly those whose members complained that brusqueness often tended into aggressiveness, that the police's desire to assert authority often took physical expression.[16] A second problem was that the professional model did not accomplish its purported mission of eliminating crime. The police, it turned out, could not fight crime with only diminished community participation; they needed neighborhood informants and eyewitnesses.[17]

Both of these shortcomings of the professional model were seen as remediable by advocates of community policing. The police could shorten the distance between themselves and urban residents by walking beats on foot, opening up substations, attending community meetings, and otherwise making themselves more open to casual contact. Police organizations would empower lower-level operatives, like patrol officers, to devise creative approaches to solving particular problems, to better incorporate the input from residents that was generated from these casual contacts.[18] Residents' satisfaction with the police would grow, and with it their desire to work cooperatively and productively with the police.

An added benefit to all this, ideally, would be the intensification of social and political ties between residents in urban neighborhoods. To deal effectively with issues of crime and disorder, residents would need to coalesce around a common understanding of these problems and a common strategy for addressing them. The dissolution of crime and disorder, then, could serve as a means by which communities could experience that sense of political efficacy that, many argue, humans need for self-fulfillment.

All of this sounds well and good. What could be wrong with exposing the police to greater citizen contact? Why not encourage community members to know their problems in common and to address them collectively? Why *not* a more responsive state, a more capable community?

But just what *is* this community of which people speak?

Three Visions of Community

The literature on community is, like the term itself, difficult to grasp. Because community often seems self-evident, the term rarely gets used with much precision.[19] As a consequence, different notions of community are deployed by commentators with little overt discussion of how those notions compare and contrast. For example, community often refers to a primarily social entity, a locus for relations of familiarity and comfort. At other times, community is extended from this social basis to include a political entity that can mobilize these relations of familiarity to protect the group's interests. This is precisely the move made by many communitarians, and by advocates of community policing. Part of the legitimacy of community policing thereby is grounded in the expectation that communities can be sources of both social comfort and political possibility. Accordingly, I seek here to distill out essential differences in approaches to community and its political potential, and to outline three approaches that receive significant endorsement. This schema does not capture *everything* that separates distinct visions of community's power and possibility, but it does isolate the most important differences.[20]

Two of these versions of community can be labeled "thick." Each sees community as an important social arena, each celebrates the benefits that ensue from communal connection. However, they differ in a critical way. In one version, community is something that is *recovered*. It is principally a repository of values that sustain a group's moral fabric. These values ostensibly provide meaning and direction for community members. The goal of community action is to acknowledge those extant values and reinforce them. In the other "thick" version, community is *discovered*. Here, there is less a sense of a fixed and essential set of values to reinforce than there is a universe of possibilities inherent in communal politics. In this vision, community members come together to create an inclusive and open-ended politics; they explore a range of political avenues in a capacious and respectful fashion. Community is thereby forged in this process of political exploration.

Each of these "thick" versions can be critiqued on both normative and practical grounds; perhaps thick connections are neither desirable nor possible. As a result, a "thin" version of community possesses great legitimacy that owes much to the central thrust of liberalism. Here, communities are seen as looser and more occasional confederations of like-minded individuals who are capable of coming and going from a range of possible such confederations. Greater emphasis is accordingly placed on the right and desire of individuals to choose from various social groups; as one's interests and life goals shift, so might one's choices of associates. Communities may well be consequential, but they are not as primordially essential as in thicker versions.

In the rest of this section, I review each of these approaches in more detail, before proceeding with an examination of whether and how they resonate on the ground.

Thick Version I: Community as Recovered

It is not by accident that discussions of community are often tinged with a degree of nostalgia, a regret that community has been lost.[21] Indeed, much of the classic sociological work on community, as noted earlier, describes modernity's ostensive power to dissolve the communal bonds that held small-scale societies together. These bonds, reinforced over time, worked in an informal yet potent fashion to organize social life in predictable ways. Even in the present day, there is a sense that in rural communities such informal normative systems still work to regulate life; they provide, in Robert Ellickson's words, "order without law."[22] Community is thus the critical crucible where this informal value structure is forged.

If these values are indeed critical to the moral fabric of a social group, it follows that community must provide structure and meaning to people's lives. This is precisely what motivates much of the communitarian critique of liberalism: the fear that an excessive focus on the individual and his/her rights implies a neglect of the sustenance that strong communities provide. This is both a sociological and a normative argument. It is sociological because it emphasizes the notion that rights are never divorced from visions of the good. As Sandel notes, we preserve the individual right to worship because we value the good that a spiritual life enables.[23] Further, such a vision of the good is never individually constructed, but rather developed through social processes. It is therefore false to treat rights as *individual* accomplishments, because they emanate from socially produced normative systems. Because these systems provide a necessary structure to our lives, they deserve protection and strengthening. Hence the normative thrust of this version of communitarianism: the need to remember and reinforce the value systems communities produce.

These values are seen as either already there or as having once existed in the not-too-distant past. For example, Robert Putnam encyclopedically documents a decline in the extent to which Americans come together in various social groups. He mourns the fact that we are "bowling alone," encasing ourselves in progressively smaller shells and losing the social capital that can only develop in collective endeavors.[24] Consequently, he and others urge a return to such collectives, to help imbue our lives with the meaning and direction that community efforts provide.[25] In the process, a lost sense of moral order can be retrieved.

Community, then, is notable primarily because of its role in the informal structuring of society through processes of value reinforcement. We are thickly connected to one another, and we would do well to acknowledge and celebrate those connections.

Such a view of community does not lack for critics. This vision is most vulnerable to the charge that it can reinforce a status quo that would inevitably diminish voices from outside the mainstream. What is typically missing from these assertions of community is any possibility that the value system might not be consensually shared. There is little or no consideration of how to resolve value conflicts that might emerge, no sense of what political mechanisms can be employed to arbitrate differing visions of the good.[26] Those who advocate such a thick notion of community cannot, as Bickford notes, help us understand "what happens when the selves constituted by different communities clash."[27] Critics thus fear that assertions of community can mostly work to reinforce value systems that are narrow and whose perpetuation advantages some members at the expense of others.[28] The norm of consensus may presume a pre-existing closeness that does not exist.[29] It is therefore not surprising to hear this vision of community derided as "moral authoritarian communitarianism,"[30] as a "cloying, unconflictual simulacrum of community."[31]

From this perspective, in Roberto Unger's words, "By its very nature, community is always on the verge of becoming oppression."[32] That is because community is necessarily exclusive; those who do not share what unites the group must be denied membership. This line between insider and outsider is often extremely consequential, marking a division between acceptable and unacceptable behavior, between those who are powerful and those who are not.[33] And this line may not always be fixed, and thus, in Bauman's words, "forever in need of vigilance, fortification, and defense."[34]

The creation of strong communities can work to increase social divisions by decreasing tolerance for those against whom the group is defined. It can also increase divisions *within* a particular group. In her classic analysis of a New England community governed by town meetings, Jane Mansbridge found that some members possessed unusual degrees of power. Typically, those who were well educated were more confident in their ability to speak publicly, and those who were longstanding members were better able to cite past practices as justification for resisting change.[35] For Morris Fiorina, the concern is with what he terms "the dark side of civic engagement," the too-common tendency for those who are especially impassioned to tilt the balance of power toward their particular concern. The more shrill and less tolerant members of a collective can thereby rule the day and prevent a more open and reasoned debate.[36]

As we will see below, those who defend liberalism suggest that a thick vision of community is illegitimately constricting. It posits an implicit image of the individual as so deeply enmeshed in communal processes that one's capacity to distance oneself from those processes seems either nonexistent or illegitimate.[37] Liberals, as Michael Walzer notes, concede the reality of socialization processes and their inescapable impact on moral consciousness, but still assert a self capable of critique. As he puts it: "Contemporary liberals are not committed to a presocial self, but only to a self capable of reflecting critically on the values that have governed its socialization."[38] For critics, a role for such analytic distance is difficult to imagine in a thick vision of community that stresses the deep impact of communally shared values.[39] And, by extension, any critique that could emerge from groups outside the mainstream might also be delegitimated. As Nancy Fraser notes in her celebrated critique of Habermas, visions of the "public sphere" have historically imputed a restrictive notion of who constitutes that public. Thus the assertion of a unitary public sphere can perpetuate the marginalization of excluded voices.[40]

But perhaps community can still be "thick" without necessarily reinforcing a fixed system of values and preferences. Perhaps community can be discovered instead of recovered.

Thick Version II: Community as Discovered

The normative impulse toward small-scale democracy heralds an active and engaged citizenry that understands and solves collective problems. Implicit here is a citizen whose self-actualization depends in part on self-determination, on the capacity to exercise influence over governance. Importantly, these exercises in political influence typically occur in the context of some sort of group. Thus, the pursuit of self-determination inevitably becomes a communal enterprise.

If the vision of community as recovered argues that values importantly predate our individuation, in the alternate thick vision, community is developed *through* political enterprises. Community is thus a process, not a thing. Barber puts it succinctly: "Far from positing community a priori, strong democratic theory understands the creation of community as one of the chief tasks of political activity in the participatory mode."[41] Such participation works best, Barber and others argue, when members approach one another in a spirit of openness and respect, and thereby enable alternate perspectives to be aired and constructively debated. Democratic action means opening oneself up to others, and thus enlarging one's perspective; politics consists of compromise, negotiation, respectful consideration of alternatives.[42] "The idea," Fraser writes, "is that through deliberation the members of the public can come to discover or create a common good. In the process of their deliberations, participants are transformed from a collection of self-seeking, private individuals into a public-spirited collectivity, capable of acting together in the common interest."[43] Further, members ideally remain indefinitely open. The goal, according to James Bohman, is less reaching some ideal end state than constructing a conversation that is able to continue.[44]

However, any debate, no matter how longstanding or open-ended, requires some sort of structure. As a consequence, writers in this tradition often emphasize presumably abstract and neutral procedures that should work to organize the political process. Thus Habermas stresses the importance of the "institutionalization of . . . procedures and conditions of communication."[45] He further notes the need for the objective evaluation of alternate political schemas in terms of more universalizable goals, such as equality and justice. That said, any set of procedures or any set of larger goals are themselves legitimate, to a significant extent, because they resonate with the less formal value structure that undergirds the group. In this fashion, as even Barber recognizes, a vision of community as recovered is implicit in a more overtly democratic vision of community as discovered.

Still, these visions *are* distinguishable by their comparative emphases on recovery and discovery, on the extent to which they advocate that our energies be directed toward reinforcing existing values or toward creating ever more inclusive and effective political groups. But this latter vision of community as discovered is no more lacking for critics than is the version of community as recovered. One issue is whether this model rests upon an accurate anthropology: are people *that* motivated to be political? Does our self-actualization demand self-determination in the political arena? Can such communitarians, as Stephen Gardbaum asks, defend "the substantive moral claim that the best life for human beings requires us to live in and as political communities"?[46]

One might also question whether politics breeds tolerance. Perhaps politics works more to entrench pre-existing positions: one enters the political arena with a particular parochial concern and defends it against all opposition.[47] This is why Fiorina worries that the more shrill voices disproportionately reverberate in political discourse.[48] In such an impassioned and embattled political theater, the development of capacious, constructive engagement seems unlikely. Associational groups may well operate according to rules that violate norms of equality and respect.[49] Empathic and open-ended political discussion might require more of us than we are able to provide, and it may be more rare than we like to believe.

Resistance to a community of discovery might thus rest in part upon a cultural predisposition to individualism, a fact of social life that advocates of a thin version of community take for granted.

The Thin Version of Community

According to critics, thick versions of community overstate our preferences for a deep connection to others or for a collaborative political project. Perhaps community cannot be trusted because it suppresses minorities/individuals, perhaps it cannot sustain the degree of political activity that some people desire. Many therefore prefer to see communities more in thin terms, as collections of individuals who sometimes come together, sometimes not.[50] Emphasis here lies on the individual and on his/her capacity and desire to choose the social groups with which he/she will affiliate. Communities thus have less claim on the moral and political grounding of individuals. Indeed, the extent to which communities form is a contingent outcome of individual choices to gather collectively for whatever purpose. Individuals can move from community to community, they can be more or less captured—socially or politically—by communities.[51] This is a vision of a "limited liability" community, in Janowitz's terms, where the range of individual connection and commitment to community is not preordained by some ostensive deep normative process, but varies from person to person, from place to place, from time to time.[52]

Such a thin version is defended by advocates of liberalism because it preserves the individual capacity to choose amongst goods. If community harmony and value reinforcement is the principal aim of social and political life, then citizens can be construed as means to ends, rather than as ends in themselves. As I noted previously, this concern is particularly heightened for those who lie outside the dominant vision of community, whose values somehow do not conform with the status quo. To assert the dominance of community in the face of such pluralism might require minorities to sacrifice their capacity to live in accordance with their preferred values. Better, therefore, to emphasize the individual's ability to choose the social groups with which to connect, to pursue the goods the individual finds most meaningful. If community can be cloying and conformity-inducing, then we must buttress the individual's capacity to break free of social bonds that are unjustifiably limiting.

None of this means, of course, that social collectives will not emerge to engage in political action. Rather, those collectives will be looser and more temporary than thick visions of community anticipate.[53] Such collectives mesh well with the thrust of pluralist politics, which envisions a political arena of contending parties. To the extent that individuals affiliate with a range of different groups in dispersed and overlapping networks, this works well to reduce political contentiousness. To quote John Ehrenberg, relying on the seminal work of Truman and of Almond and Verba: "Multiple memberships and overlapping loyalties drive toward compromise and integration, and a moderate liberal democracy is best able to satisfy the broad range of interests generated in civil society without large-scale political disruption."[54] Such a democracy, John Rawls asserts, should structure politics in the most general way possible, it should avoid endorsing a particular moral schema. This, he hopes, will yield procedures that are widely acceptable and commodious, and that rely upon publicly defensible reasons for their perpetuation.[55] In other words, thinking of community in thin terms might be the best way to ensure an open-ended, polyvocal, and inclusive political process.

For advocates of liberalism and pluralism, then, community is not inconsequential, but its power must necessarily be limited. We must ensure an individual's opportunity to embrace alternatives and to engage and disengage with various collectives as priorities and allegiances shift. Such a robust sense of individualism is further entrenched in American culture in a range of ways, including the importance of the individual as a consumer in capitalist society.[56] Indeed, communitarians often bemoan the hegemony of individualism because it diminishes meaningful community connections.[57] But skepticism of the excesses of community, and the dominance of images of individuals as property owners and marketized consumers in capitalist society, ensures that a thin version of community will necessarily compete for legitimacy with thicker versions.

Normative theorists, then, construct competing arguments about whether and how communities should form, about the strength of the connections that can and should exist between members of social groups, about the way a politics built around community should unfold. One way to pursue this debate is to try and determine the grounds upon which one can legitimate one of these visions as superior to another.

I push in a different direction, and ask instead whether any of these visions resonate with residents in urban neighborhoods. Indeed, a common critique of normative theorists of community is that they pay little attention to the actual meanings of community.[58] To assess the potential of community means to overcome this problem. What *do* people say when you ask them about their visions of urban community? To ask this question is to inquire whether a project like community policing can possess any traction. Because justifications for community policing typically rely, implicitly or explicitly, on thick visions of communal life, it is necessary to ascertain whether those visions possess any significance for the citizenry. If urban residents do not share the assumptions that underlie community policing, then the project may well not succeed. Residents may possess a different sensibility about the weight-bearing capacity of community.

Community on the Ground: "Workable if not Intense"

It is one thing to make normative assertions about what community is or should be, or to build policy around the idea that a community can be an effective partner with state agencies like the police. But it is something else entirely to expect that these assertions about community will possess any resonance or utility in the community itself. What if residents in urban neighborhoods do not themselves share the normative vision of a thick community? What if they neither possess nor desire the degree of connection that community policing presumes? What do urban residents want from a community—something thick or thin, or nothing at all?

Each of the citizens interviewed was asked to talk about community, to explain what the term meant. They were asked to discuss the various social groups to which they belonged, and whether any of them rose to the status of a community. In so doing, they explained what, to them, constituted a community. They were also asked to talk about the neighborhoods in which they lived, and to assess whether those neighborhoods warranted the label of community. Recall that community policing, like other moves to maximize local democracy, rests upon an assumption that a neighborhood can, in fact, be a community. It was thus essential to determine whether, in practice, the two are conflated by urban residents. As respondents expanded on their answers, a remarkably consistent set of ideas about neighborhood as community emerged. People of different incomes, races, and ages all returned to much the same vision of community. Notably, this vision does not conform neatly to the thick and thin versions elaborated by theorists, but lies somewhere in between.

All of the respondents displayed a desire for their neighborhood to possess a sense of community: it clearly persists as a term that feels good. However, none of the respondents described community as a repository of fundamental values or as a forum for the elaboration of a consensually generated political program. In this way, they provided little hope that a thick version of community can underwrite projects of neighborhood governance. But their vision was not entirely thin, either. There was instead a sense that living together in a neighborhood implied a connection that needed to be established and reinforced. This sense of connection, however, did not derive from values or political programs but from the basic needs for ontological and physical security. Urban residents wanted to know who their neighbors were and to exist in relations of basic friendliness and cooperation. They wanted to be able to predict what their neighbors were going to do on a day-to-day basis and to rely upon them when in need. Such basic knowledge worked to reassure residents that they could manage their daily affairs without disruption and live in their homes without tension. As Chloe, a Westside resident, summarized the point, the goal is for "workable if not intense" relations with one's neighbors.

INTERVIEWER: Would you call this neighborhood a community?

CHLOE: Um. . . . In some senses of the word, I would. Though, there's . . . like most cosmopolitan cities, there's a sense of people being off on their own, and somewhat disconnected, because people seem to—in cities very much—maintain their own life inside their property. But what does make it a community is the fact that people are satisfied with their neighborhood, and stay here almost permanently, and have harmony with their neighbors even if there isn't as much exchange as there used to be in the olden days, and the fact that they're content enough to stay and have workable if not intense relations with their neighbors.

INTERVIEWER: So you would describe most of the relationships here between neighbors as "workable" rather than "intense"?

CHLOE: Yeah, maybe workable is the word. Not super connected, but very friendly when there's a need to connect.

As Chloe suggests, she is not "super connected" to her neighbors, and she expresses no regret about this reality; she does not desire a thicker connection. What does explain her happiness with her neighborhood is its permanence and predictability, and her attendant certainty that her relations with neighbors are "workable" if any issues do emerge. She readily acknowledges that the connections in her neighborhood could be "more usable or visible or something." But, she says further, "I'm not sure anybody feels the need for that. If we felt the need, I'm sure we would do it, but I think everybody's probably satisfied with things the way they are."

This vision of "workable if not intense" relations was one echoed almost universally by those interviewed. What matters is neither a strong sense of shared values nor the capacity to act collectively in the political arena. Rather, residents hope to know and rely upon neighbors, to feel that their basic needs for security can be met in the immediate environment. Said Tracey, a Midlands resident:

TRACEY: For instance, our street, we all talk to each other. And we help each other out if we need help. The guy across the street is the guy who drove me to the hospital when I had my baby. It's very friendly. And there are schools in the area that make it feel like we are supporting our children. I like some of the stores here, they are not like the malls and stuff, they are just nice small little stores, they are comfortable. Some of the people know us by name, so it feels good that way. Yeah, that kind of feeling.

INTERVIEWER: Knowing each other.

TRACEY: Yeah, and being aware of each other.

These twin desires—to know one's neighbors and to be able to rely upon them for basic assistance—were repeated, in various ways, by a wide range of those interviewed. In explaining why he would, in fact, characterize his neighborhood as a community, a Beachside resident, Bob, said: "I think that if somebody ran into trouble, neighbors would help. We are interested in the welfare of each other, because we like each other. In that . . . in part what describes community to me is it's a group of people who have something in common and will help each other." Sally, a Midlands resident, explained how nice it is to simply wave at her neighbor as she walks home from work, and to be able to help out if needed: "One time [neighbor's] car got stuck down at the Water Taxi, and so she flagged me down and said, 'Hey, could you give me a ride?' And I felt really good being able to help out, and it didn't take a lot of my day." Respondents expressed happiness that they could borrow ingredients for recipes, share dishes of food, reciprocate pet-watching services during vacations.

These priorities on familiarity and reliability directly influenced how respondents understood their responsibilities to each other. To the extent that they worked to establish communal relations in their neighborhoods, it was not because they affirmed moral values or developed a politics, but because they made themselves known and assisted others as they were able. The following four respondents developed variations on this theme:

Nora, Eastside resident: I think for me community would be defined as knowing my neighbors by name—not necessarily knowing their personalities, but at least feeling comfortable to stop and chat with them a little bit: knowing what kind of car they drive, and do they have kids, and you know, what are the names of their children.

Barb, Centralia resident: I think the first responsibility is to just get to know who your neighbors are. Get to know their names, whether they work all day, and let them know things about you, just to build relationships. It doesn't have to be like your neighbor comes over every other day for coffee, but just so you know who they are, you know that in that house [name] lives there and she works all day and she usually gets home around 5, you know, that kind of thing. Just on your street, just get to know each other and let them know you.

Carrie, Midlands resident: My attitude is normally that it's great if you can get friendly with your neighbors if you don't feel like you need to be that close to them as long as you are friendly or cordial and that kind of thing, and that you sort of watch out for your house and watch out for their houses. And you hope that your neighbors do the same for you when you are not around.

Kristi, Eastside resident: I think it's important that you communicate with your neighbors—know who they are on a first name basis. Know when are you here and when are you not here. Do you need me to watch things. Some people will consider it being nosy, but I think it's considered being smart, because you know what's going on in your neighborhood and then you can see when things aren't quite right—somebody's hanging out that shouldn't be hanging out.

If there are common values uniting these residents, these revolve around security, not some deeper morality.[59] Residents seek comfort through knowing that their neighbors are people with whom they can minimally relate and upon whom they can occasionally rely. They want to be able to understand and predict what will happen daily in their locale. They want to know when others are coming home from work, where in the neighborhood they can acquire a tool or emergency babysitting, whether they can rely on others to ensure proper maintenance of homes and yards. In this predictable and reliable landscape, residents can expect minimal disruption and can easily identify those ruptures when they do occur. You can, as Kristi said, "see when things aren't quite right," and respond accordingly.

Not surprisingly, the lack of this predictability and reliability was noted by those who lived in neighborhoods where they appeared to be absent. As Roland, a young resident of Centralia, noted about his neighborhood, which is dominated by multifamily housing: "Community is something you read about in the newspapers." For him, because of the transient nature of his neighborhood's population, durable relations were impossible. His neighborhood, he noted, was merely a place through which people passed, "a foothold before they get into a community."

These and other impediments to the development of sustained, solid relations in neighborhoods will receive attention in chapter 2. For now, the point to stress is that those interviewed never described urban community as a crucible for either moral grounding or exercises in political self-determination. The thick versions of community, in other words, possessed no resonance; neighborhood as community does not appear to mean either a deep repository of shared values or a means by which people act politically. But it would also be inaccurate to describe the hegemonic sense of community as thin, because there is an invariant realization that the reality of propinquity breeds a desire for a lasting and predictable connection. The implicit recognition that they are anchored together in place means that residents feel a need to establish ties that are enduring, not contingent. Values are implicitly emphasized, but these relate to issues of ontological and physical security: a desire to be secure in one's surroundings, a hope that others will be helpful if asked, a simple interest in being able to neighbor if one wishes. Residents want to know that no threats exist, and recognize the need for some level of connection to ensure that sense of security.

What, then, are the conditions that enable neighborhoods to be secure in the manner that residents most desire?

The Building Blocks of Community

Because residents commonly expressed a desire for predictability and security, they were fairly consistent in enumerating those factors that make that desire a reality. Three factors, in particular, were regularly cited as important: property ownership and its associated virtues of pride and permanence; children and the means by which they serve to connect neighbors, including through local schools; and various activities that bring people outside and thus position them to engage in casual neighboring.

Residents regularly correlated the possibility of predictable relations with length of residence. If they and their neighbors were in place for some time, then they could more easily develop a greater understanding of each other's daily patterns.[60] Residents thus nearly uniformly favored neighbors who were owner-occupiers of their homes rather than renters. Ownership not only implied permanence, but also pride and care. Property owners were seen as more likely to maintain their homes in good condition. This helps increase the collective belief that the future is predictable. Residents took comfort from their neighbors' home improvement projects, just as they feared the consequences of a rental property in their midst:

Sarah, Eastside resident: Well, I like my neighbors. Most of the people that live here, you know, if you came into my yard you could see that fence we put in this spring. People are working on their yards, working on their houses. There is a lot of, well, "I'll get this little house and I'll fix it up" kind of attitude. So the people that are moving into the neighborhood are really making big improvement. I like that, I think that's nice.

Richard, Westside resident: I see us as in the right hands, the neighborhood is getting better. But it doesn't take much to have one house that would be a rental and be slowly deteriorating and I think it would really pull the neighborhood down. To sell it, it's pride in ownership, or hopefully there should be.

These respondents echoed a common theme: the importance of property ownership for establishing the framework upon which ontological security can be built. Those who owned property were seen as more constructive community builders; their commitment to their domicile implied a degree of permanence and pride that would maximize the collective sense of predictability. A lack of pride in place would, by contrast, loosen the skein of neighborhood connections. Those respondents who lived in neighborhoods with a mixture of owner-occupied homes and rentals were especially fixated on the damage the latter could do to their sense of security.[61]

Property ownership was the most commonly cited variable for establishing the conditions for relations of security. What helped matters further, many respondents suggested, was if those homeowners were also raising children. Parents, they suggested, were particularly attuned to the dynamics of the neighborhood, ever aware of those families which also included children and more inclined to be proactive in establishing ties. As Stephanie, a Midlands resident, put it: "People tend to get associated with each other more quickly, I think, through their kids. People are interested in who their kids are playing with, and the kids who come to your yard and play." In a similar vein, Sally, a Westside resident without kids, commenting on her lack of strong connection to her neighborhood, said: "I think that maybe if I was married and had kids then I might feel a little bit of a more, I would have more connections maybe in the community itself, in this area, I would have more reason to have interactions with some other people in this area." Care for kids, then, as with care for home, increased neighbors' connections with one another.

By extension, many respondents cited the importance of a local school as a means by which a sense of community could be sustained. This point was typically made by those who believed they lacked such a school, or who believed that the policy of busing students throughout the city negatively impacted community building. As Lane, a neighborhood activist who lives in Westside, summarized it:

INTERVIEWER: But your sense has been, since you've been there, that there has been a bit more . . . a bit less of that kind of community feeling than you would like?

LANE: Oh, not a bit less, a lot less. I mean, because it's, it's just part of the modern pace of life and a lot of different changes that happened in the '60s and '70s where you had—and into the '80s—where you had a lot of white flight to the suburbs. And then people doing things like not putting kids into the community schools, in Seattle particularly, that's been, you know some people see that as, you know, something necessary and then you can also look at it as a real . . . that was a large factor that I personally think contributed to the breakdown of the sense of community, because people . . .

INTERVIEWER: The schooling issue?

LANE: Yeah, the schooling issue.

Betty, also quite active in local organizations, made much the same point:

What is strange is getting a lot of kids bused in. And it's not always for ethnic mix-up either. It's just that the kids get bused in. And if the school district and the city saw the value in making each school as good as it could be, you wouldn't need to bus kids. Kids could choose if they wanted to go to another school, but the neighborhoods would be stronger because of that.

For these respondents, as for many others, schools were an important means by which families could become more anchored, and could thereby establish a stronger sense of neighborhood and community. The lack of a strong neighborhood school was a notable source of frustration for those who sought to build community.

If property and procreation were considered critical means for establishing ties in neighborhoods, so was the simple activity of being outside. If residents desired a basic degree of connection, then opportunities for such casual neighboring were obviously enhanced if they regularly saw one another. Many respondents mentioned particular neighbors who made a point of taking care of their yards or of walking pets regularly, and thus served as a conduit for community building. Gardening was repeatedly mentioned as a critical means for neighboring, as elaborated most eloquently here by Marsha, a Westside resident:

Tracy was a great motivator, our first year, she worked at her front yard, and it's hard to see how beautiful her yard is without thinking maybe I could do this. So she really inspired me. Last year, our neighbor, the dahlia neighbor, ripped up their front yard and made theirs a cottage garden, so again I would say his wife is one of the influential people in the neighborhood because she has really ignited a spark of gardening. And gardening, we are outside, we are pulling things apart and saying hey I have too much of this, it grew too big, would you like some. So we ended up starting to talk, so I think that's been a community building activity, gardening.

Any means by which neighbors were brought out of their homes and placed in a position to engage in casual conversations served as an important means by which neighbors could establish stable and secure (if rarely intimate) relations.

Property, children, and regular encounters were thus recognized as essential ingredients for predictable neighborly relations. Residents who possessed these in their neighborhoods expressed appreciation for them; those who lacked them noted their absence. Not surprisingly (and this is an issue to explored in more depth in chapter 2), the presence or absence of these factors correlates with social class. Neighborhoods of greater economic advantage are more likely to be composed of stable homeowners, to possess a local school of high caliber, and to be sufficiently safe to enable casual outside neighboring. As economic advantage declines, each of these potential neighborhood realities becomes less likely. As chapter 2 makes plain, this is quite consequential when neighborhoods do choose to engage in political action.

Conclusion: Neither Thick nor Thin

Many possess tremendous hopes for community. As a source of moral reinforcement or as a forum for creating social and political capital, community is often seen as a fundamental component of a good life. And, to the extent that localized democracy is heralded as an ideal, this community should be based on propinquity. Such localized units are defined in geographic terms, in no small part because similarly situated people face similar problems from such possible sources as pollution and crime. The high hopes for community rest upon a normative geography of strong neighborly relations. These hopes undergird community policing, which presumes that local ties can be the foundation for projects to reduce crime and disorder.

Neighborly relations are, in fact, desired by those we interviewed, but not for the reasons touted by devotees of robust community. Residents look to their neighbors not to collectively develop a web of life-anchoring values nor to develop an enlarged and effective politics. Rather, they seek to know one another well enough to go about their daily business with minimal disruption. These are relations of basic familiarity rather than intimacy, of casual contacts rather than political discovery. They want simply to locate themselves within their milieu, to feel secure in their knowledge of those who surround them, to see their neighbors as reliable. Sharing space means sharing some collective vulnerability, so residents seek ties of just enough strength to minimize that vulnerability. This suggests that thick visions of communal togetherness and collective governance are misguided; projects, like community policing, that place significant political weight upon communities are unlikely to succeed.

Yet it is a mistake to describe these idealized neighborly relations as thin, because the reality of contiguity does imply a connectedness that informs residents' sense of a desired community. Although they recognized that the demands of community must be attenuated in a modern, fast-paced, individualized world, they still wanted their neighbors to recognize their shared spatial interdependence and to participate in community enough to reinforce a basic level of connectedness. They expressed regular appreciation for a sense of community that provided them reassurances that they would be assisted in need and protected from threat.

Even if this vision of neighborhood connectedness does not meet the criteria of a thick vision of community, it might not conflict with variants of projects like community policing. If neighborhood residents desire security, and if they experience some sense of connection to one another, then collective action against threats to their safety and property might well occur. Perhaps such a vision of community—as a minimal web of relations focused on security—could still be effective in developing a political program to ensure that security. Perhaps community could bear some of the weight of communal governance presumed by advocates of community policing. It is toward a more explicit analysis of the *political* capacity of community, particularly in terms of relations with the police, that we must now turn our attention.

The Political Status of Community

Kristi, a middle-aged, middle-class woman who resides in Eastside, is immensely disturbed by a young man who lives down the block. The young man lives with his mother, and generates much neighborhood angst. He drives cars very fast down the street; he works noisily on those cars late at night; he welcomes many short-term visitors to his home, breeding suspicion that he is engaged in drug deals. At length, Kristi discusses the young man's actions, how they disconcert her, and how she is responding. Asked at one point whether she feels influential in her neighborhood, she responds with an enthusiastic "Yup!" When asked why, she says: "Because I refuse to turn my head and stick my head in the sand and pretend that it's okay with what he's doing and it's not. And I will continue to fight, because I own property here and he does not. And he has chased six neighbors away in the past six years. They just get fed up with it and can't handle it any more. I'm about there myself." Note the speed with which Kristi shifts from determination to despair. Indeed, later in the interview she says, "I don't know. I'll probably end up moving." Six years with the problem have taken a toll, and she admits to hopelessness. Discussions with other residents who face similar challenges reveal that she is not alone.

I showed in chapter 1 that urban residents, when asked to define community, discussed its social mode, as a pattern of neighboring that can ensure feelings of security. Almost never did anyone suggest that community should be primarily about political activity. However, even if community is not primarily about politics, it is plausible that it could still sometimes serve that purpose—particularly with issues of crime and disorder. If neighbors are concerned about security, and if they have at least somewhat familiar relations, they might collectively thwart a perceived threat. They might be able to bear the political weight that community policing would place upon them. Might their shared concerns and connections be sufficient grounds for collective action?

In the interviews, residents discussed security, and whether their community could increase it. Like Kristi, most residents expressed considerable pessimism about community as a vehicle to address matters of security. This doubt was voiced more loudly where economic distress and visible criminality were more commonplace, but it was a theme in middle-class neighborhoods as well. A surprisingly common set of concerns emerged around the political viability of community. These concerns underlay their pessimism.

This despair obviously contradicts the rosy scenario often painted for small-scale democracy. The devolution of political authority should empower those closest to on-the-ground dynamics. Because neighborhood residents know most about phenomena like crime in their midst, they are best equipped to devise strategies to reduce it. As a small-scale collective, they are also plausibly more likely than a larger group to engage in consensual, capacious decision making. Such devolution is thereby arguably consonant with democracy. It is also frequently a component of projects consistent with neo-liberalism. Though it is sometimes used in an overly broad fashion,[1] neo-liberalism describes efforts to reduce the role of the centralized state in economic and social affairs. Instead, social life is to be structured primarily by market forces. State projects are delegated to either private enterprises or voluntary institutions.[2] Community often emerges in neo-liberal governance as the recipient of some of the responsibilities for charity and security the state now eschews.[3] Community policing exemplifies neo-liberal efforts to devolve power in the name of increasing localized self-determination.

My analysis leads me to add to the chorus of voices raising doubts about the ultimate sensibility of such devolution.[4] While one can applaud increases in a citizen's sense of political agency, foregrounding community does not often accomplish this. It is easier to share the doubt expressed by the citizens of West Seattle about the prospects for political transformation through community.

I explore this pessimism in this chapter. I move through three main sections. In the first, I review the four factors the residents cited most frequently to explain the political lightness of community: individualism, heterogeneity, transience, and fear. Although these four issues emerged frequently, the interviews demonstrated that the severity of these challenges correlates with economic standing. Neighborhoods that are more disadvantaged suffer from these impediments more than do their more comfortable counterparts. Constraints face all neighborhoods that embrace self-governance, but they weigh more heavily on some.

The significance of the unevenness of these constraints becomes more obvious in the second section, in which I review issues of representativeness and cultural capital. I recount resident commentary about how leadership in neighborhood groups is often not representative of the group as a whole, in part because of the many requirements for success. The bureaucratic hurdles placed before communal groups are high and complex, and only certain citizens possess the ability to surmount them. As a result, particular residents emerge as leaders, residents who do not always reflect the composition of the group.

Thus, the capacity for success might be geographically uneven, because some neighborhoods house more residents with the necessary cultural capital. In the third section I review the widespread sense that wealthier communities are indeed better able to make localized democracy work for them. And even where less advantaged communities do achieve success, it results from concentrating their attention largely upon small-scale matters. I therefore consider whether community governance can significantly impact the dynamics that most strongly shape those neighborhoods.

All of this generates significant doubts about the political project of devolved authority. This provides cause again to wonder whether community is unbearably light. This will become clearer as we review each of the four most frequently cited impediments to communal political action.

The Impediments to Communal Politics

"People Like to Be Left Alone": Individualism and the Demands of Community

Urban residents want neighborly connections of just enough strength to guarantee predictability and reliability. This quest for security is understandable, given the spatial interdependence that neighbors share. That these connections are not deeper is due in part to the valorization of individual autonomy. Just as there is a cultural warmth attached to community connection, so is there a resentment of the nosy neighbor. As much as Americans romanticize community, they also treasure freedom and autonomy. Between these two extremes, connections of basic familiarity seem a sensible middle position, a balance between group togetherness and individual freedom.

The American valorization of individualism is a frequent source of communitarian complaint.[5] To the extent that Americans primarily engage themselves in projects of self-fulfillment or material advantage, they may well diminish the moral fabric of their collective. While this complaint can be criticized for naively yearning after a vanished time when notions of community were allegedly more regnant,[6] it was one echoed by many of those interviewed. Recall Andrew, quoted at the beginning of chapter 1, waxing nostalgic about how there "was such an involvement" in the neighborhoods of his past, how "everybody knew everybody" in his days of yore.

However often the respondents indulged in such nostalgic cravings, they more commonly suggested that this state of relations is no longer possible, or even desirable. They may want neighborly connections, but not at the expense of their independence. Said Walter, a resident of a beach neighborhood, when asked whether he lived in what he considered a community: "It's kind of a community of individualists [laughter]. Um, people, I, you know, it's a community . . . the beach focus is what really kind of, I think, glues people together. Whether they're friends with each other, that's another matter." Later in his interview, he indicated that political organizing was quite rare in his area, only occurring once or twice in his nearly twenty-five years of residence. In his neighborhood, close connection and formal organization were not the norm.

While only Walter used the term "community of individualists," his description of the loose connections that bind him to his neighbors was common. Overly tight relations would inevitably be too constraining, the respondents indicated, an unwanted intrusion into one's business. A desire for individual autonomy limited the depth of neighborhood interactions.

Individualism created barriers to communal involvement.[7] This was frequently noted by community organizers; they said their neighbors were otherwise engaged, and thus often reluctant to participate. Here's Richard, a neighborhood watch leader in a middle-class community in Westside:

I think generally people like to be left alone, to their own devices. They feel that if they involve themselves in the community, there's going to be a commitment that they don't want to give. I guess there's always a sense of, "Oh, it's going to cost me." Well, really it's not going to cost you. All I'm asking you to do is to say hello to your neighbors or keep an eye on the street if you're out there. That's all I need you to do. I'm not asking for you to cough up a hundred dollars a month to pay for private security or anything. Just be a little bit more involved and just making people feel like they are a part of the block or a part of the community and just say hello and make people feel comfortable and secure in the knowledge that, "Hey, I can count on you to at least watch my house."

Andrew, also a watch leader in a middle-class neighborhood in Midlands, made much the same point:

Well, I think it's an attitude of like, "I don't really want to get involved." And I have to be guilty. I do that too. I don't want to be involved. I'm busy. I have my family to take care of. I'm self-sufficient. I know what I want, I know where I'm going. I'll throw my arms around my family first, before I actually go out and do something for somebody else. I think everybody is sort of guilty—it's just an American thing generally.

The notion that a desire for individualism was rooted in American culture was expressed most directly by Nelson, a resident of Blufftop, who was raised in a small African village. In his childhood, he knew well all of those who surrounded him. The contrast with neighboring in the United States was striking:

But in this country, what I've seen is this culturally kind of individualism, which is good in some cases but is also not good in some cases. And there is no . . . there is no . . . it is very hard for you to live here and know your next door neighbor in that sense.

This emphasis on individualism was expressed invariably, and volubly, by those who tried to lead neighborhood activities. Many of these activists recognized that efforts to increase involvement could not be pushed very far before their neighbors would refuse to be drawn in further. Even if there is considerable enthusiasm at one point to act as a group, this often diminishes over time. Here is Kristi again:

Every once in a while, down here at the community center, they have community meetings for the area. And because every neighborhood, every block has a different complaint, you hear so many. And a lot of them complain about the drug activity. This area is just nasty with drug activity. But when we go down to the meetings, we get to see other people in other neighborhoods. "Yeah, yeah, yeah, we need to get together and talk about this." But you never do. Nobody ever finds the time. And lately, well, we've been . . . well, the last one was over a month ago—all new faces. There's some of the old faces, but mostly they're all new faces. And you know, they're people, "Oh, we're new into the neighborhood, we're going to make a change." Well, yeah, go ahead, give it a try. When you figure out the magic words, let me know, because I'm going to do it in my neighborhood!

Efforts to create cohesive neighborhood political organizations often lacked the right "magic words" to surmount the predisposition to individualism. For many residents, concerns for oneself and one's immediate family impeded collective action. Often, the demands of daily life were too overwhelming to allow time for neighborhood politicking. Said Frieda, a retired middle-class homemaker, when asked why she did not join her neighborhood group:

Because I'm tired. You know, it gets to be too much, you just get so, it's just meetings all the time, so everything you join you have to commit your time to, and right now I love to garden and I go up at the church, and weed around the flowers there, and just, and help out a neighbor or two around here, so. When you are in all these clubs you don't have time for that. So that's why I have just. . . . As it is it seems I'm still, it seems like I'm active in too many things as it is.

Many activists empathized with this complaint, particularly when describing the circumstances of those less economically advantaged. If Frieda, retired and economically comfortable, struggles to find time to be active in neighborhood matters, imagine the challenges for those who struggle to avoid penury. Lane, the head of a community crime prevention organization, recognized the difficulty for many citizens in West Seattle of engaging in efforts to increase security:

Some of them frequently work two or three jobs where they don't have a neat nine-to-five schedule, where they have the rest of the time to be involved in the community and to meet other people. I mean, it's harder for them if they are working a different schedule that is not in that neat nine-to-five category to be active and involved in neighborhood meetings.

Jean, another activist in a mixed-income neighborhood in Eastside, echoed the theme: "I think you start losing some of the community, because people are, perhaps, concentrating on working to get by."

In short, the pursuit of individual aims constrains organizing in all neighborhoods, but impacts some more severely. In neighborhoods of concentrated disadvantage, organizers face the Sisyphean task of recruiting participants for whom political action seems a luxury that time and circumstances do not permit.

These citizens, then, made clear that neighborhood cohesion is frustrated as individuals engage in matters other than community building. Activists seek the "magic words" to coalesce their neighborhoods, but recognize the intractability of the cultural predisposition to individualism. Individual pursuits, even basic economic security, often trump localized political activity. A similar challenge to community political action is posed by heterogeneity, which also possesses an uneven geography, and hence an uneven effect on community organizing.

"A Huge Barrier": The Constraint of Heterogeneity

One of the common critiques of community is its inherent exclusivity. To define a group necessarily requires a demarcation between those who belong and those who do not. For many commentators, this demarcation can be reactive and defensive, an attempt to attain a group purity that reinforces social differences.[8] These differences, many note, are increasingly inscribed directly in the landscape, as the well-advantaged seek protection from those against whom they define themselves.[9] Indeed, one can argue, following McConnell,[10] that the more local the group, the greater the likelihood of homogeneity. In well-defended communities, a more inclusive polity cannot form. Walled off from those they fear, the prosperous elude responsibility for the wider public. If community is invoked as a defense of these practices, then its pursuit becomes a suspect political project.

For citizens in West Seattle, difference was indeed experienced as a constraint on the pursuit of community. Yet what they expressed was not so much a defensive posture against those who were different as a frustration with how difference interfered with simple communication and deeper forms of connection. It is, of course, entirely possible that the respondents stifled expressions of discomfort with social groups different from themselves, for fear of disapprobation. However, many of them, particularly those who expressed genuine interest in cross-group organizing, found heterogeneity a complicated matter. And the severity of the challenge that heterogeneity presents is not felt uniformly across space. Because the degree of demographic difference is uneven across urban space—and aligns directly with economic class—its effects on political organizing are also uneven.

Where people sought to unite their neighbors, they were frustrated when linguistic and other cultural differences asserted themselves.[11]

Residents did not indicate that they believed that difference needed to be exorcised from the community. Instead, they suggested that simple lines of communication could not be opened. The following excerpt, for example, comes from Jean, a middle-aged, middle-income white woman who is married to an African American and is the president of her neighborhood association. She seems genuinely eager to engage her neighbors of Asian and Latino backgrounds, but does not quite know how:

I think one of the major things I've seen—and again, I've lived here probably seven years but really didn't get a good feel for the neighborhood until maybe the last three years—language is a huge barrier. Language is a huge barrier. Kids learn the English language, but they don't necessarily communicate with their parents what their schools are asking them to communicate, so their parents aren't exactly aware of what's going on.

Throughout her interview, Jean circled back to the intractable issue of communication. She mentioned the need to print up flyers for community meetings in different languages, to try to diversify the gathering. Left unaddressed was how communication would proceed if representatives from all the neighborhood's linguistic groups were present. Others active in community-based efforts to combat crime frequently mentioned the need to communicate with those who spoke different languages. They regularly stressed, for instance, the importance of advertising the fact that the police's emergency call system employs operators who speak different languages, so that no one would hesitate to call the police.

The fact that neighborhood activities were invariably conducted in English and generally presumed a basic familiarity with the operations of state agencies such as the police understandably limited the involvement of groups without those capacities. As a consequence, those populations were largely invisible to many activists. As Dee, the head of an Eastside neighborhood watch group, put it:

I guess I'd have to guess that the population is—in my neighborhood—is predominantly white, but I would also say that there's a really good chance that that is not true and that, because I just don't . . . the folks that are not white, in my experience, tend not to get involved in the community action committees or block watch groups or . . . for whatever reason.

The non-whites in her neighborhood are thus not entirely visible to this community leader. Further, she cannot explain their absence.

Part of the problem, perhaps, is that cultural differences are more than just linguistic. Nora, an Eastside resident active in neighborhood watch activities, expressed her frustration with trying to make connections with her neighbors of Asian heritage:

I think . . . I think there are some cultural barriers with some of the Asian families that make it difficult to communicate with them, or them to feel . . . I don't know that they really understand what block watch is all about . . . so, being able to have an interpreter that would be able to explain that to them. And then there are just, you know, I don't know . . . I have personal hang-ups about . . . about the roles that women from Asian countries get put into, and so—mom always being at home cooking and cleaning is really difficult, because I would love to go over and, you know, and chat or meet them, and there's not really that opportunity. And when I've tried that with some of the men in the family, I really don't get much of a response from them, and I assume that it is because I'm female, but, you know, I don't know that that's true. It might just be them. But I do think that cultural differences like that make it challenging.

The intractability of cultural differences was noted in nearly every community where heterogeneity existed. Such heterogeneity was most pronounced in Blufftop, the large-scale public housing facility in Midlands. It was not uncommon for up to six different language interpreters to attend public meetings there, a noteworthy attempt to ensure that all could be engaged. However, this did little to break down the barriers between these groups. Said Dan and Heather, an American Indian couple of long residence:

DAN: No, they stick to their own groups, the only people that mix with each other are. . . .
HEATHER: The Vietnamese always stick together, the Samoans always stick together with their own group. Filipinos stick with their own groups. These people, Somali stick with their own, and this other, all the Blacks got their own, the American Blacks have their own.

For some, disjunctures between cultures are not just challenges, but sources of ongoing tension. In one particularly distressed Centralia neighborhood, relations between African Americans and Latinos were often antagonistic, as Darla and Jocelyn, two young black females, explained:

INTERVIEWER: So you said the Mexicans and the blacks on [street name] don't get along?

DARLA: No.

INTERVIEWER: Why is that?

JOCELYN: Cause they think that sometimes they . . . the blacks . . . have like music out-side they get mad, but when they be having their music out we don't care. They be having their music. . . .

DARLA: They think that we're taking over . . . but we lived here first. They just start coming cause everybody moved out.

Such clashes also took a violent form in this neighborhood. Both residents and the police traced several homicides in the area to battles between ethnically based gangs. A clear narrative of what prompted this violence was elusive; battles over control of the drug trade were often cited, as were disputes over women. But these groups were ethnic-based, and provided the most striking reminder that cultural distinctions were significant.

What clouds this picture further is that outsiders cannot recognize that most groups possess internal divisions, as well. Nelson expressed discomfort that all Africans in Blufftop were yoked together, despite their considerable internal variation:

NELSON: No, within East Africans, maybe, there is some kind of underneath. . . . It's not something that you can see if you are not a person who is a part of it. But it never comes to affect anything or create any disturbance or anything, but maybe they are not . . . a little bit they might not feel happy about each other. In that sense, yeah. But, of course, like an East African and a Southeast Asian, no problem, there was nothing.

INTERVIEWER: The internal tension. . . . And is that between people from Somalia and people from Ethiopia and. . . .

NELSON: No, I think maybe within people from Ethiopia. Like for example, there is the Eritrean and Ethiopians which is traditionally, or culturally, connected in some way. And because of a war they had back home, created an atmosphere kind of thing, and I think even within Somalia, maybe, there is some kind of problem that they have in their country that might play some role over here.

Similarly, Meg, an African-American resident of Centralia, explained what happened to her when she chastised other blacks for stealing the tools of a white neighbor. After she told them their stealing reflected poorly on "all of us," she was strongly criticized:

Oh, boy! Why did I say that? They called me a sell-out. I was called everything but a child of God. A traitor . . . to the race! I said, "Wait a minute! Hold up! I don't play that!

Wrong is wrong—I don't care what color the person is. When it's wrong, it's wrong! Don't do that to me."

Differences within and across groups, then, impede the creation of political ties in urban communities. Given its correlation with economic disadvantage, heterogeneity's impact on organizing is unevenly felt. At a later point in her interview, Jean described this challenge in her efforts to organize in her neighborhood:

But because I think of the lower- to middle-income families and the diversity, it's hard to recognize it as one community. Lots of cultural differences and people tend to stay in their own little separate entities. So it's hard to bring all the communities together to make one big community.

Heterogeneity and class combine to make elusive Jean's work at cohesive political community.

The struggle to establish connections across and between different linguistic, cultural, and ethnic groups was regularly cited by residents as a principal impediment to the community cohesion they thought was necessary for political capability. In the next section I consider transience, which was viewed as similarly detrimental.

"Just Disgusting Filth": The Taint of Transience

Urban residents desire predictable and reliable neighbors. Community is principally about a security that is maximized when one knows one's neighbors' time-space patterns and feels comfortable asking them for occasional assistance. It is therefore hardly surprising that permanence was valued and transience disapproved. For residents, predictable relations were more likely if their neighbors were residents of long standing. With exposure to one another over time came greater understanding.

Residents thus favored homeowners as neighbors. Ownership implied permanence, pride, and care. Owners were said to be more likely to maintain their homes. This increased residents' feelings that the future was predictable. Renters, by contrast, were seen as too transient and too lackadaisical to contribute to a collective sense of ontological security. This was expressed most colorfully by Carrie, a Midlands resident:

Yeah. Oh, yeah. You can look at our yards. The people that own and care, take care of our yards—you can tell. And, I'm not saying we go into each other's houses every time something goes on, but if you do go into one of our houses, you'll see, we care. It's picked up, it's cleaned up. You go into a renter's house, it's just disgusting filth. Just disgusting.

Hal, a renter himself, residing in Centralia, expressed much the same sentiment:

INTERVIEWER: How would you say this neighborhood is different from the ones right around it?

HAL: Well, if you went across [street name] and got further over in that area, they're homeowners, and it's a lot more pride in there, you know. Over here there's no pride in nothing.

Property ownership is connected to ontological security. The commitment people make to their homes implies a broader commitment to locale. Renters lack any such commitment, and thus were seen by respondents as culturally suspect and politically unreliable.[12]

If a neighborhood consisted of mostly owner-occupied homes, its cohesion and predictability were maximized. By contrast, a neighborhood with many rentals was undesirable. Renters were understood as unconnected to place. Just passing through, they could not help glue a community together. By extension, neither could they be expected to join in any neighborhood efforts oriented toward political ends. Hal, a resident of one Centralia neighborhood dominated by multifamily housing, was asked to describe the sense of community in his neighborhood.

HAL: Well, what community? I mean, who do you got? You've got [neighborhood leader], and he's good when he's here, but when he's not, it's just so, what do we do now? It's that way. There is no, "yeah, we'll handle it." There's no community coalition or anything. There is a few people, but with a few people gone out of it, it doesn't work. . . . And none of us renters are involved in it.

INTERVIEWER: None of the renters?

HAL: No. I mean, why should we be? Because we're renters! Why should we be involved in anything? We're paying these people to live here. It's not like we own any of this. There's no pride in renting something.

Renters, it is commonly believed, care about neither their property nor their neighborhood; they cannot be relied upon for political muscle. This was noted not just by homeowners, but as the above quote illustrates, by renters themselves. This was even more eloquently expressed by an Eastside teenager, Roland, an unusually articulate analyst of the conditions of his community and his role in it:

ROLAND: In a nicer world, I would be responsible for: keeping my neighborhood clean, which I do; trying to keep it safe, which I don't; and you know, and putting in . . . trying to give back to the community, by helping, like weed, or do things without an ulterior motive. Now, the fact is, is this: I'm just like everybody else. I go to school, I study, I go to work, I do everything I feel like I should be doing to be successful. And, as soon as I get enough money, or as soon as the opportunity presents itself, I'm out of here. So, this isn't exactly like my number one choice for being here, so I'm not exactly the right person to be talking about who should be doing something for the community.

INTERVIEWER: So, because you feel that way, are you saying that keeps you from investing in the neighborhood?

ROLAND: Yeah. I wouldn't . . . you know, you don't want to invest in a car that you know is going to break down sooner or later. Sooner or later, I'm going to be leaving, so I'm not going to put money into it. I'm not going to put in my time. I don't know, it makes me feel like a sucker, but, you know, that's just the way it is.

Given the hegemonic notion that renters are disconnected from place, it is unsurprising to learn that they were sometimes ignored in efforts to mobilize community, particularly around issues of crime. Chloe, a Westside resident, said:

Well, I know the block watch captain—maybe this is policy, but—he does the roster, and he has all the names and phone numbers of the owners, so that you can call each other if someone is prowling your car or whatever, and he said that the way it stands, we don't do the rental people. And I don't know if that's for security reasons, cause they kind of come and go, or . . . I'm not really sure. They do come and go. So, yeah, and we just don't know them as well. They kind of keep to themselves. I think they're not as plugged into that sense of community as the rest of us are as home-owners.

Home ownership thus assumes a political significance, an indicator of the commitment that makes cohesion possible.[13] Those without that status are not political equals, because they presumably cannot forge a minimal degree of connection.

For non-renters and renters alike, property ownership was an important marker. It follows that neighborhoods with a disproportionate share of renters would understand themselves as less able to mobilize politically.[14] This puts such neighborhoods at a significant disadvantage when political projects are devolved to lower spatial scales; those communities in greatest need of organizing are the least able to accomplish it.

"It's None of My F'ing Business": The Constraint of Fear

Both casual neighboring and focused political action require some base level of familiarity and comfort. This comfort is, in turn, contingent upon conditions of security, such that residents can commingle in space. If one experiences the immediate environment as dangerous, one might well withdraw from visibility. This imperils community relations. One's potential vulnerability is heightened further when one raises the stakes and takes concerted action against neighbors suspected of criminal wrongdoing. Suppose a resident believes that drug sales are occurring in the neighborhood and grows frustrated with the ancillary activities often associated with it—excessive traffic, noise, possible violence, and petty thefts. To confront those who appear to be involved is to take a significant risk. One is exposed particularly to the possibility of retaliation, a factor that deters many from acting in the first place. Rob, a Centralia resident, acted anyway, but understood why fear afflicted his neighbors. When discussing his quite visible actions—organizing block walks, directly confronting suspected drug dealers—he was asked to contrast himself with his neighbors:

INTERVIEWER: Are there other neighbors who wouldn't do anything at all about these problems?

ROB: Oh, yeah.

INTERVIEWER: Why do you think that is?

ROB: Fear.

INTERVIEWER: Fear of what?

ROB: Retribution by the police. Or by the drug dealers. Or by the landlords. And it all happens, all of the above. You complain too much and you get in trouble.

The belief that complaining led to trouble was commonly cited as an impediment to community building. Given the possibility of danger, better to simply withdraw, a point made clearly by Roland. Asked to describe his responses to wrongdoing that he might witness, Roland said:

ROLAND: If I walk by and I see something which I may think to be shady, I'm not just going to run to the nearest phone that I see.

INTERVIEWER: Just turn a blind eye to it?

ROLAND: Pretty much, yeah. Like I said, nobody's bothering me, nobody's doing anything to me. It's not my problem.

INTERVIEWER: So, why is that? 'Cause you're not looking for trouble?

ROLAND: It's not my concern. It's between the shady person and the shady goings-
on, and the people who are there who have a problem with it. I'm not saying, let
me see . . . I'm not that callous. If I saw someone getting beat up, I'd call the
police, if I saw something like that happen. If I saw, you know, if I saw someone
breaking into a house, and I'm across the street by a phone, I would call the police.
Let's say, if I saw somebody steal a car . . . I don't know, that's sort of shady, I might
just turn a blind eye to that, unless I knew the person with the car. I'm not totally
callous, but if there's something where it's not violent, and there's not something
that I can immediately do, I'm not going to go out minding to it.

The possibility that avoidance was the prudent course of action was
echoed by Kelly, a middle-aged woman who lived near a home she
believed hosted drug dealing. When asked her response to the noise and
speeding cars that she associated with the alleged drug sales, she said:

KELLY: For a long time, I would call the police, but I've had several neighbors tell me
it's none of my f'ing business, so, if it happens in front of their house, I don't see
nothing going on. But let it come in front of mine, I'm on the phone.

INTERVIEWER: Do you ever go out there and confront somebody?

KELLY: No, wouldn't do that. Nope. Not in this day and age. I've got kids. I'm going
to see them graduate.

INTERVIEWER: So it's for fear of what might happen?

KELLY: Yup. Definite fear. Definite.

Both Roland and Kelly express a conflicted moral position. On the one
hand, each recognizes an obligation to their communities, a recognition that they *should* respond to criminality. Yet they also recognize the
possibility of danger. Better, then, to minimize their individual vulnerability.

Fear thus makes it difficult for many to address directly instances of
actual or suspected criminality. But it also works to deter casual neighboring outside. Those who live in neighborhoods that they perceive to
be unsafe—because of activity around locations of suspected drug sales,
because of speeding cars, because of occasional physical threats—express
reluctance to spend too much time outside. This point was made most
eloquently by Carrie, who lives a block away from a residence out of
which she believes drugs are sold. The high-speed traffic on her block
limits her neighborly connections:

Like I said, my fairy tale would be to live in a cul-de-sac where everybody took care of their yards, and the kids all got along, and were able to play and didn't have to worry about speeding cars. And I know of neighborhoods like that—I've get friends in those kinds of neighborhoods. So, I guess I envy them, because that's what they have. In order to make this a community, we would have to block off the end of a block so the cars can't race, or put in speed bumps or something, so people aren't afraid to let their kids cross the street once in a while. I mean, my kids know that once they hit the gate, if they cross it they're dead meat, because I've told them and told them and told them, if they ever get hit by a car, there's going to be nothing I can do for them. Nothing. Because they go that fast. So, to make it a better community . . . I mean, I think our little block is a good community, we just don't like the traffic, we don't like the drugs. That's what makes it not a community. We can't go in the middle of the street and carry our cup of coffee and talk, because it's only seconds and someone's speeding by.

Marsha, a resident of a Westside neighborhood with a similar problem, was asked what she thought would create a stronger community. She laughed and said, "Gates, big high brick walls, and guards."

Fear significantly dampens many residents' willingness to engage with others to lessen their exposure to criminality. Indeed, some pointed to the capacity of those engaged in wrongdoing to use fear to minimize community meddling. Recall Kristi, the woman quoted at the beginning of this chapter who is irritated by the loud antics and frequent visitors of her young neighbor. Besides these ongoing practices, she said, her neighbor uses scare tactics to protect himself. He intimidates by confronting those he believes lodge complaints with the police. As Kristi summarized it:

KRISTI: Well, the community's afraid of him. Everybody's afraid of him. I'm afraid of him. But I'm not going to let him know that. Because that is his leverage—he knows people are afraid of him.

INTERVIEWER: Why are people afraid of him?

KRISTI: Because he's very aggressive, and very angry, and he explodes at the littlest thing, verbally, all the time. He's always swearing at his mom, he punches holes in her front door.

In the face of such intimidation, the sensible strategy, as Kristi subsequently noted, would be to coalesce as a united front. As she put it: "If all of us went over there and talked to him and looked him right in the face and said, 'Hey, we will not tolerate this anymore,' it probably wouldn't happen so much."

But happen it does. The paradox of fear, as with individualism, heterogeneity, and transience, is that it works its greatest damage on those neighborhoods that most need to overcome it.[15] To expect a crime-ridden neighborhood to organize to increase security is to ask its residents to confront a fear that is often extremely well-founded. If fear indeed impedes community organizing, its effects are not evenly distributed. Once again, it becomes evident that the move to devolve political authority to community does not necessarily empower those neighborhoods who suffer the constraints imposed by economic disadvantage.[16]

"It Seems to Be the Same Three to Five People": Representativeness and Cultural Capital

At the time of his interview, Mark was part of a loose coalition of community activists in Centralia. He joined weekly block walks and attended most meetings and neighborhood events. He did more than his share, but he expressed consternation that his group's numbers did not grow. Recruitment efforts invariably failed. As he put it, "And that gets back to the whole community. It seems to be the same three to five people that organize events and this and that."

Mark isolated an intractable problem with the political mobilization of community—the issue of voice. Can anyone ever legitimately claim the authority to speak for "the community"? Upon what basis can such a claim be sustained? Might such claims be inaccurate, or unavoidably partial? Might those who act as the voice of the community be articulating a narrow, parochial view, and thus obscuring a more diverse reality? Are those who are voiceless thereby further disadvantaged?[17]

The significance of this concern increases when one acknowledges the challenge to political cohesion presented by the four factors outlined above. If a community group lacks a wide membership, then its representatives will likely lack an accurate perception of the political desires of all the residents.

This problem was widely acknowledged by those interviewed, including those who were active. In particular, the respondents noted a persistent tendency for the more active to possess similar demographic characteristics. Leaders tended to be Caucasian, comparatively well educated, middle- or upper-income, and owners of property. As Rob said about his quite diverse Centralia community: "But this is a telling thing about this thing here is that, the surface of this community is homogenized. The people that have a voice are homogenized." Barb, a black resident of the same community noted that she was usually the lone person of color at the local Chamber of Commerce meeting, even though the business community was diverse. As she said, "There are a lot of minority and immigrant business owners but they, like I said they sort of stick to themselves." This dynamic made it difficult for any community leader to claim an ability to represent the group.

Yet many of those interviewed argued that strong leadership was necessary for communal success. Significant bureaucratic hurdles face any neighborhood group that wants services from the police department or any other state agency. Therefore, many suggested, those who were well educated and institutionally deft needed to emerge from the pack. Otherwise, any group efforts were likely to fail. To receive resources required a long, time-intensive effort to first understand and then mobilize the process. For the crime prevention activist Lane, this explained why the well-educated emerged as the dominant voice. Success required attending frequent meetings, chatting up key officials, and being articulate in public forums. You needed, in her words, to be "really vocal in the neighborhood process. You make those ties and people listen to you. And if you're not part of that, then it's a whole lot harder to get people to take your complaints or to take you seriously." As she noted, if you possessed cultural or linguistic differences from this mainstream, you likely could not enter the discourse.

Thus, one had to be able to "work the system," a skill invariably associated with higher levels of education. The process of mobilizing a community's concerns to ensure a hearing from government agencies was understood as difficult and time-consuming. Further, one needed to be acquainted with various types of discourse that were considered effective. As Rob put it: "It's the people who have the power and that are articulate, that can talk the talk and walk the walk." He knew this well, from personal experience. A high school graduate and a carpenter, he was nonetheless the leader of his local group. He was pressed by his neighbors to apply for a grant, but he found the process humiliating. He was not, he said, given a positive reception when he turned in the grant. He said, "I was dissed on it, because I had filled it out in carpenter pencil. But I had written a really good mission statement and something else. It didn't meet their standard, because of that. It seemed kind of funky." He resented how his comparatively limited education was used, in his perspective, as a factor against him and his neighborhood.

Even though many applauded strong leadership, others resented the lack of representativeness that often resulted. The dividing line between the active and the inactive was often a line of tension. A sense of a lack of adequate representation frustrated those who felt excluded as a consequence. As Meg, the African American quoted earlier, put it:

That's another thing I don't like: people coming in making decisions that haven't been there that long, don't get a consensus from the neighborhood of, "is this what they want" or "is this going to make the neighborhood better?" Then when they do something and we don't agree with it, their attitude is, "So what, I don't care. We're going to do it anyway."

Her grievance is thus with others who do not seem to accommodate her views and yet presume to speak for her. She wants a stronger sense of representation. But those whom she criticizes argue in response that, without their work, nothing would get accomplished. Rob, whom Meg was implicitly criticizing, complains about the lack of initiative displayed by his neighbors:

And when I'm gone and stuff, we allowed . . . I mean, they . . . when I was gone another time for three months and I came back, they allowed people . . . they didn't call the police—people don't call the police, they don't do stuff! It's just like so frustrating! "You allowed them to take over the neighborhood while I was gone!" And they'll do it that quick!

The struggle to create neighborhood political associations generates questions about the representativeness of the groups that emerge. Leaders argue that they must act independently in the face of community inaction or else little will be accomplished; those who feel left behind complain about the diminution of their voice. Even Rob acknowledged that his actions might be the cause of resentment, a feeling he betrays at the end of the following comment: "As a leader, I pick what I want, and then we lead them into that. And they like it. You know, pretty much."

One could argue that strong leadership was a necessary evil in neighborhood politics, even if it compromised broad-based representation. But this persisted as a significant source of tension. As noted, those who felt excluded often resented those who spoke in their name. Conversely, residents who felt overburdened with responsibilities resented those who absented themselves. Andrew, a neighborhood watch leader, described his feelings when, shortly after he took ownership of his house, he found himself directly involved with a neighbor who hosted boisterous parties. He recalled his thoughts during a shouting match with that neighbor:

ANDREW: I don't know, I wish there was more of a united front. It seemed like it was just me and this guy [neighbor] that were really doing something about it.

INTERVIEWER: So, a united front means in terms of other neighbors?

ANDREW: I think there should have been more involvement with at least some of these people here [points to houses in immediate vicinity].

INTERVIEWER: So they just didn't want to get involved.

ANDREW: Yeah. It would have been nice to put on a united front and just say, "We're all concerned," rather than just have one person or two people basically shouldering the responsibility and taking care of it. I felt like I was there all alone. I felt very alone. The middle of the day and it was just me and this guy. And I thought, "What am I doing here? I just moved here! This has been a problem for a year before I got here. Why didn't you guys do something? Or, why didn't you come up to me and say, 'Hey, let's try to get these jerks out of here.'" So, I thought it was above and beyond, personally.

The division between the involved and the uninvolved can thus become a significant fault line. Besides being an activist, Lane lived down the street from an older man, Vern, who was the source of long-standing complaint. Vern parked several junked cars in his yard, allowed visitors to come and go all through the night, and threatened any who complained about him to the police. Some neighbors actively assisted the police, including videotaping the activity at his house to help provide evidence of drug dealing. The active neighbors bore Vern's threats, and came to resent their uninvolved neighbors. As Lane described it:

LANE: And so the people who really wanted to get it cleaned up and get the drug house gone felt some anger and resentment to the people who didn't.

INTERVIEWER: Because they felt more at risk?

LANE: Yeah. They felt . . . because they then became the targets for the harassment. Everything from car windows being broken to poison being put out for pets to . . . to having houses broken in and stuff stolen, and the people who tended to have that happen . . . well, not tended, in every instance the people who had that happen were people who were reporting and wanted to try to get something done. So they felt like they were being singled out because they were the ones willing to stick their necks out, so they resented the people who weren't willing to do that, because those people were the ones who were going to benefit, when it finally got straightened out. So I mean, if . . . all so many issues get divided in just one little block.

The problem of representativeness resists resolution, and persists as an understandable source of tension when neighborhoods mobilize.

Even if the possession of cultural capital qualifies some residents uniquely to work the system to the neighborhood's benefit, the assertion of their authority can cause resentment from those who feel left out. Conversely, leaders may feel unsupported and unappreciated when their neighbors passively sit back. Tensions can thereby divide "just one little block."

The importance of cultural capital recalls again the significance of class, a point many emphasized when they compared their neighborhoods to others nearby.

Scale, Class, and Success

One of the strongest arguments against devolving political authority is that it disproportionately benefits well-advantaged communities. Education and political involvement are positively correlated with economic class,[18] and metropolitan areas are economically segregated.[19] Thus, if communities assume increased responsibilities for self-governance, then some neighborhoods will likely prosper, potentially at the expense of others. Affluent neighborhoods may be able to ensure that social problems like pollution and crime are kept out of their locales.[20] If they can so succeed, they necessarily foist their problems onto those neighborhoods less capable of organizing.

Thus, the devolution of authority might crimp a wider sense of public responsibility. As Donald Moon poses the question: "If collective control is to proceed through small, decentralized communities, then do we not sacrifice the ability to shape our social world as a whole?"[21] This philosophical question acquires greater urgency through Nina Eliasoph's sociological analysis. Eliasoph found that the various volunteer groups she studied typically restricted their sociological imagination to occlude discussions of broader political processes, the better to minimize internal dissent and to maximize their sense of efficacy. The groups accordingly discouraged broader debates about the causes of the social issues upon which they concentrated.[22]

The devolution of authority, then, arguably restricts the political scope and sociological imagination of local groups, in ways that disproportionately benefit affluent neighborhoods. This line of argument was endorsed by many of the citizens interviewed. For example, there was a widespread sense that success in neighborhood political action was linked to economic standing. This complaint was expressed loudest by residents in neighborhoods of disadvantage, who complained that their needs were not taken as seriously as those of the more well-heeled. But such a view was widespread. Witness, for example, the following excerpt from Bob, an upper-middle-class respondent who said he lives in a neighborhood with a strong sense of community:

INTERVIEWER: Do you think that some of these groups, or one of any of these groups, is more powerful or influential in the neighborhood than others?

BOB: Yeah, sure. The more affluent. The top of the . . . the view areas. What we used to call the lunatic fringe.

INTERVIEWER: Why is that? Why did you call them that?

BOB: We have very few very rich areas and very few very poor areas in West Seattle, as a whole. And it was just a name that [friend] and I came up with. We'd call them the lunatic fringe, because they're . . . some were more vocal politically and socially than others. And this includes the view property up here and the beach property down there, on the edges of West Seattle.

INTERVIEWER: Why do you think that was?

BOB: It had to do with economic station. And they were more vocal.

INTERVIEWER: In neighborhood politics?

BOB: Yeah.

Rebecca, a middle-class resident of Midlands, also argued that political clout is tied to income and education. Those who make noise, she said, "are the ones who have money. And they have property. And they have education. And they can articulate what they need better than people that fear the government or fear the cops or immigrants or whatever."

The opportunity cost of pursuing assistance from the city—the hurdles to be overcome, the rules for how requests are made—was consistently seen as extremely high. For many groups, the obstacle course of bureaucracy discouraged their involvement. And when those efforts that did emerge received only minimal state response, when the weight of disadvantage seemed impossible to overcome, the temptation to withdraw became strong. Said Patty, a resident of a disadvantaged neighborhood in Centralia:

Sometimes I despair. But I think there's a glimmer of hope there. I'm losing some good neighbors that I interacted with a lot, and that I had a sense of community with, and they're going, once again. And the other neighbors have been here for seven years—they had their tires slashed and they're now talking about moving.

Residents of poorer neighborhoods suggest that the devolution of authority punishes them for their comparative difficulties in mobilizing governance effectively; they see little impact from whatever efforts they mount. This lack of effectiveness might also be tied to the matters upon which they concentrate. Neighborhood groups typically focus on matters in their immediate environment, such as poor lighting or inadequate sanitation. These are obvious manifestations of the problems that beset them, and imminently sensible challenges to tackle. Similarly, those who were focused upon creating more cohesion attempted to build community. Such efforts included putting up signs at the borders of neighborhoods to promote communal identity, or erecting kiosks for posting notices of community events. Lane, for one, possessed high hopes for such efforts:

We're really working to try to rebuild that sense of community. We're doing, we put the signs up that are around the area that clearly define, you are now here and entering our neighborhood, and we are working on kiosks that are going to be community information—they are going to be these big covered areas where people can put up anything to do with a community event or activity, post things, open for anyone in the community to post things. We're trying to just kind of reconnect people to what else is going on.

However, it is hard to see such efforts amounting to much. Given the challenges to communal governance, the occasional kiosk seems rather inconsequential.[23] Indeed, the devotion of communal energies to issues of such limited scope calls to mind Eliasoph's conclusions about volunteer groups: that they limit their sociological imagination to those matters over which they feel some agency, the better to maintain enough optimism to persevere.[24] Indeed, at another moment in her interview, Lane confessed that "If you look at the scope of it, it seems overwhelming." Perhaps this helps explain her focus on kiosks, an immediate alteration to the landscape that could provide a hopeful sign of community. Yet it is hard to assess such efforts as particularly potent means to enable more effective communal politics.

Conclusion: Going through Hell

The high hopes many possess for neighborhood governance are easy to understand. Urban residents do know their areas better than government bureaucrats; political self-determination is a laudable pursuit; cohesive communities can provide immense support. For these reasons, projects like community policing seem entirely legitimate and defensible.

But this presumes that communities are not unbearably light, that they can support significant political weight. This chapter makes plain that obstacles to communal cohesion are numerous and consequential. Further, these obstacles place their largest burden upon those communities in greatest need. Problems may persist, even worsen, if devolution deepens, as increasingly minimized resources are distributed unevenly. Because economic health is connected to political capability, disadvantaged neighborhoods stand to suffer, not prosper, from increased expectations of communal governance. In the trenchant words of Schlozman, Verba, and Brady, "It is naive to expect the institutions of civil society to be the magic remedy to overcome the class-based participatory deficit, for the proposed cure contains the seeds of the malady."[25] Or, in Lawrence Sherman's words, placing high expectations on poor neighborhoods to solve their crime problem "may amount to throwing people overboard and then letting them design their own life preserver."[26]

This line of complaint was engaged by Patty, a resident of a disadvantaged Centralia neighborhood whose needs she believed should be a priority. She argued that expectations for her community were simply too high, that they could not meet all the requirements for obtaining the resources they needed. As she put it, "I mean, say if there was some city funding, or some money available that we didn't have to go through hell to get, there could be a lot of projects around here for people." She believed that devolving authority to her community was ultimately a raw deal, a means by which comparative disadvantage was maintained.

Patty's argument is made more sensible if one understands the political opportunity structure her neighborhood faces. Community policing presumes not only that communities possess political capacity, but that their voices will be heard by state agencies like the police. To discern whether that is the case, one must turn attention to the police. One must consider how the police relate to community and how they structure their work. Such an investigation requires us, again, both to assess conflicting normative visions and to compare those visions to matters on the ground.

Elusive Legitimacy: Subservient, Separate, or Generative?

It is the regular monthly meeting of a neighborhood council in Eastside. The neighborhood in question covers about a single square mile. Its residents are mostly middle-class homeowners, although there are also many multifamily dwellings. The racial composition of the neighborhood is majority white, but a sizeable minority population exists, comprising primarily Latinos and Asians.

In its meetings, the council addresses a fairly typical set of concerns: speeding cars on residential streets; possible negative externalities from a homeless shelter at a church; blighted homes; occasional graffiti. Because public safety concerns invariably form part of each meeting's agenda, a representative of the Seattle Police Department usually attends. The officer is usually included as a standing item on the agenda, typically to summarize recent crime events and to solicit questions and input from the residents.

On this night, the SPD representative is a lieutenant whose responsibilities include supervision of the precinct's Community Policing Team. After describing significant recent crimes in the area, the lieutenant asks for questions. One woman asks about a possible recent spate of car thefts from her block and wonders whether the lieutenant can confirm this statistically. The lieutenant claims ignorance, citing the fact that he only recently assumed his post. He does, however, reinforce the idea that the neighborhood is an attractive spot for car thieves. As an aside, he tries to explain this phenomenon in terms of the large number of "unsupervised youth" in the area.

This comment prompts a question. Asks one resident: "What happens to these kids when they are caught stealing a car?" The lieutenant leans into to his microphone, and in a stern voice says, "I'm going to tell you the truth, and it's going to irritate everyone in this room." He says that essentially nothing will happen to these kids, because the youth detention facility will not jail overnight a car thief or any other non-violent offender. As a result, a car thief will "never see a minute" in jail until after the fifteenth arrest, when he might get ten days. The lieutenant describes this as "one of the weak links in the system." The only sound in response is a muttering in seeming support of the lieutenant's indignation.

Community Policing and the Ideal of Police Responsiveness

The legitimacy of community policing rests in significant part upon the long-treasured ideal of localized democracy. To self-actualize, citizens should come together and exert meaningful influence over the policy matters that affect them. Community policing can be legitimated as such an opportunity for urban residents to act in their collectively defined best interest. As we have seen, various impediments can render this vision illusory for urban neighborhoods. But even if neighborhoods could develop greater political capability, they would still need to exert sway over a state agency like the police. In other words, the police would need to integrate public input into their practices. They would need, to some extent, to render themselves subservient to the citizens they police.

The police, like all state agencies, stand in fundamental connection to the citizenry; they exist to provide service to the population. But, as with all state agencies, the precise nature of the relationship between police officers and the citizens they serve is not so straightforward. As with the political role of community, this is a matter of both normative theory and sociological reality. There is normative confusion about how state and society should relate, and there are significant on-the-ground dynamics that shape how these relations play out. In this chapter and the next, I seek to make this complex reality more clear.

Much of this complexity is attributable to different idealized models for state-society relations, each of which possesses tremendous legitimacy.[1] For instance, the general narrative of democracy suggests that state agencies like the police should be responsive to public input; they should do as they are directed by the citizenry. In such fashion, the state should understand itself as primarily *subservient* to its constituents.

However, the narrative of liberalism suggests that state agencies must possess a necessary autonomy from public input, the better to protect the rights of all citizens against the possibility of unjust majoritarianism. The state should strive toward some degree of neutrality, loyal primarily to the abstract rule of law. Otherwise, the rights and needs of the more marginalized will likely be threatened. For these reasons, the state must understand itself as importantly *separate* from society.

Yet respect for the rule of law is not the only means by which such state separation can occur. State actors often consider themselves possessors of unique bases of knowledge and authority. They thereby distinguish themselves from those they govern and assert that distinction in their daily practice. The police, for example, embrace the image of professional crime fighters, and they often expect deference to their unique authority. Separation here dons a professional guise and significantly structures state-society relations.

But subservience and separation do not exhaust the possibilities for the state-society relation. One could also argue that the state is *generative* of community. That is because the state, with its policies, helps determine the central characteristics of communities. Further, the state only understands "community" in particular ways, through the routines and epistemologies state actors use to filter public input. State actors also often characterize their work in terms of an overarching moralism that situates both state and society on a transcendental plane. This is particularly true of police officers, who thereby construct community as an entity in need of the protection that only their virtuous vanguard can provide.

The lieutenant's performance at the neighborhood council meeting displays all these modes of relation. To an extent, the lieutenant is being subservient. He takes time to come to the meeting, he provides information he thinks the residents will find useful, he opens himself to questions and discussion. He reinforces the implicit assumption that, as a public servant, he must make himself available to the citizenry and engage them on matters of public safety.

This nod to subservience is, however, not all that characterizes the lieutenant's behavior. He also positions himself as separate. He sits at the front of the room, in full uniform, and speaks in an authoritative voice. Behind him stands the majesty of the law, the rights it protects, the duties it imposes. He acts as if his authority is unquestionable. This sense of autonomous authority is exacerbated by the emotional and rhetorical force of his utterance. Already an intimidating force—bedecked as he is with uniform, badge, and a belt full of coercive tools—the lieutenant imposes himself further with his presentational style. He physically and symbolically sets himself apart from the public.

Yet the performance is also generative. The lieutenant invokes implicit notions of both the community and the larger moral terrain upon which he situates his work. Community, in other words, is not separate from the state; rather, it is something that state actors, like the lieutenant, produce in their work. They render society sensible through certain organizational routines and moral architectures. In the lieutenant's case, it is his moral architecture that is most evident. Note the manner by which it leads him to assume a commonality in the group about his viewpoint. He says it will "irritate everyone in the room" because he believes everyone agrees that car thieves should be tracked down and removed to jail. He implicitly suggests that anyone who disagrees is suspect. The lieutenant's moral understanding renders the community a set of passive, like-minded good people who deserve better protection from the criminal justice system.

Thus, even though there were at this meeting elements of police responsiveness to the public and a sense in which the police were open to accountability, the lieutenant was not especially subservient. He used his institutional stature and his emotional rhetoric to short-circuit rather than promote debate about crime control policy. This was therefore not an instance of a vibrant project of deliberative democracy. Even though the lieutenant likely considers himself responsive to the citizenry, he does not approach the gathering as an open, respectful dialogue between similarly situated people. Rather, he constructs himself as society's protector and asserts his unique authority to quell a moral plague that troubles the otherwise peaceful community.

The lieutenant's performance is something of an overdrawn example, but it represents a pattern that was invariably repeated at police-community forums. Almost never did a police official seriously interrogate any suggestions from the public that the organization reorient its priorities or practices. Recall another lieutenant's performance at the Weed and Seed meeting described in the introduction. In that case, the lieutenant less loudly but no less effectively controlled the conversation by deflecting recommendations from the public. His use of these strategies may not have been entirely conscious, but it was effective nonetheless. In an interview, however, a precinct captain did acknowledge his desire to retain tight control over an upcoming community forum. If he surrendered any control, he suggested, he would get his "ass lit up." Similarly, after one such public meeting, another precinct captain expressed pleasure with how the meeting went: there was little discussion after his presentation and thus little need to respond to public concern. Such lack of enthusiasm for meaningful engagement with citizens obviously stands in opposition to the normative, democratic thrust of community policing.

In this chapter I illustrate how the police invoke different under-standings of their relationship to the community and how these do not necessarily cohere. Most of all, I show how the police understand them-selves as importantly separate from society. Of the three paths to state legitimacy outlined above, the police favor a self-understanding that emphasizes their autonomy. However, this is justified and perpetuated in terms other than those envisioned by liberal theorists. Although they do see themselves as neutral enforcers of an abstract, objective law that is ideally applied equally to all citizens, officers more robustly build a self-construction as members of a politically embattled institution whose unique base of expertise needs protection from the uninformed meddling of biased community activists. From this perspective, the police must stand separate from society because otherwise they will be enmeshed in political machinations that will negatively affect them. Officers repeatedly state a fear of being sacrificed on the altar of "poli-tics," and thus seek to stiff-arm public participation to avoid that fate.

A complete explanation for why this is the case must wait until chap-ter 4. For now, the task is to understand how the police pursue and balance the drives to be subservient, separate, and generative.

The Police-Community Connection: A Normative Approach

Even if concerns about the political capacity of community could be allayed, it remains uncertain how any communal group can and should interact with a state agency such as the police. What should be the balance of power between a community group and the police? Should the community direct what the police do? Or should the community play second fiddle, acting only via police directive? Or is a genuine partnership even possible, one in which the balance of power is equilibrated? How to strike a balance between subservience, separation, and generativity?

There is both normative and sociological heft to each of these approaches to the state-society connection. Each possesses legitimacy, and each influences interactions between communal groups and the police. It is therefore difficult to assess just how communities *should* interact with state agencies like the police. In this section, I review each of these three approaches in a bit more depth, and outline how each shapes how one might construct an ideal balance of power between community groups and the police.

Subservient?

Iris Marion Young states succinctly the case for citizen influence over the policy-making process: "The normative legitimacy of a democratic decision depends on the degree to which those affected by it have been included in the decision-making processes and have had the opportunity to influence the outcomes."[2] Implicit in this normative formulation is an active citizen who deserves a place at the table at which policy is generated. Without so respecting the agency of its citizens, a state is not legitimate, according to democratic theorists like Young. A state must ensure that its dictates possess the imprimatur of the people it serves; it must respect the capacity for self-determination that inheres in each citizen.

This is particularly critical when one considers the coercive power state agencies possess, none more obviously than the police. As noted by such important theorists as Weber, Durkheim, and Habermas, coercion is an insufficient basis for rule.[3] Exercises of coercive power must invoke some greater good, some communally supported ideal. Otherwise, power is nakedly exposed and loses legitimacy. Citizen control of state agencies that exercise coercive power is thus critical to democratic states' quests for legitimacy.[4]

Such an assumption helps legitimate community policing. Recall that community policing efforts developed in response to tensions between the professionalized police and urban communities. Professional police were seen as too aloof and too often aggressive, orbiting somehow above the everyday concerns of citizens. Such citizens were not recognized as possessing a legitimate need to direct police operations. Rather, they were adjuncts to police operations, and their input was narrowly channeled by the dictates of professional practice. By contrast, legitimations of community policing are full of language of "co-production" and of "partnerships," language that heralds a citizenry active in the construction of police policy. An engaged and active citizenry can exercise self-determination through shaping the police's use of the state's coercive power.

The language of democracy thus reinforces the legitimacy of community policing; the police need to be responsive to the will of the community. Such an active citizenry is also endorsed by theorists of crime and social control who recognize the fundamentally communal nature of deviant activity.[5] If crime and disorder develop from communally based social patterns, then these patterns must be addressed. Such a logic, for example, drives various restorative justice programs.[6] The goal here is to respond to deviance by re-establishing communal connections. Deviant behavior is, by definition, a violation of communal norms. The goal in response could be to re-establish those norms and to reincorporate the offender into the fold. Implicit here is a notion of community members as effective agents who assume responsibility to reinforce bonds frayed by deviance.

In sum, a strong version of community power suggests that communal agents are primarily responsible for ensuring social control. They do this by reestablishing connections to those who deviate and by assuming a muscular role in overseeing the formal social control efforts of the police and other relevant state agencies. From this view, the police must understand themselves as importantly subservient to the community. They must avail themselves of public input, enable citizens to help construct police policy, allow citizen groups to evaluate their performance. Without such responsiveness, the police's quest for legitimacy might well suffer. There is thus a push from many quarters to open up the police to oversight from the citizenry, through such vehicles as police review boards.[7] At the neighborhood level, this could mean partnerships that are more genuinely shared projects of governance, not police-run exercises that marginalize community input. In these and other ways, the police are seen as within the parameters of citizen oversight.

Separate?

This push toward subservience, however, faces resistance from those who fear the potential misdirections of an overly responsive state agency. Might a communal group be swayed by an unrepresentative and particularly passionate minority? Might it be motivated by its opposition to some other communal group that is perhaps not as well organized? If so, does a subservient state become inadvertently an unjust one by favoring one group's interests at the expense of another's? Might a minority group be unfairly disadvantaged in the process?

These concerns about the passions and partialities of communal political activity reinforce the liberal emphasis on a politically neutral state whose sway over civil society must face limits. These limits are made real through the creation and protection of various civil and political rights. These rights should prevent the state's intrusion into such matters as expression, assembly, and worship. A state thereby guarantees individuals and groups the chance to coalesce as they please, to pursue a range of political projects, to construct freely the "good life" they cherish. For liberals, individuals need autonomy to pursue their vision of the good. The state should thus be minimalist and neutral: minimalist, because it should rarely restrict one's freedom; neutral, because it should not promote its own version of the good to the possible exclusion of others. In important ways, then, the state needs to be separate from communal groups, constantly wary of how the pursuit of a parochial community agenda might lead to the usurpation of the legitimate rights of others.

This liberal protection of rights is not, of course, necessarily in conflict with the quest for democracy. Quite the opposite: civil liberties are often legitimated precisely because they provide the political space in which groups in civil society can construct and debate alternatives for state policy. Without such protections, the political agency of the citizenry cannot be realized. However, there are critical moments where these two approaches to legitimating the state may well come into conflict.[8] What to do when communities protest the construction of halfway houses for convicted sex offenders who are due to be released from prison? Should the state respond to the wishes of such communities, or protect the legally defined rights of ex-offenders? How to deal with the issue of jury nullification? Should juries strictly follow the letter of the law, or should they be empowered to actually nullify the law in question?

Political liberalism often emphasizes the procedures through which political discussion should ensue. The goal is to ensure that public discussion is fair, orderly, and protective of a plurality of groups. Thus, John Rawls exalts "public reason" as the mechanism for adjudicating political conflict.[9] The goal is to try to remove from public discussion as many metaphysical and moral questions as possible, to narrow the debate to those issues around which agreement is possible. As Jeffrey Isaac, Matthew Filner, and Jason Bivins summarize it, "Public reason is typically abstract and juridical; it is dispassionate and 'rational,' that is, oriented toward the uncoerced agreement of deliberative interlocutors; it is expressed in a way that is accessible to others in spite of their particular identities."[10] This should minimize the divisiveness of a plural society. But whatever calm is produced in the process comes at the price of a more open discussion. In this way, the state is to be shielded from the responsibility of adjudicating claims that originate in such moral discourses as religion. It is to be responsive to only those policy advocates that can articulate a case with an abstract logic embraceable by all. Constructed like this, liberalism restricts the extent of democratic debate and renders some citizen demands as beyond the political pale.

The extent of state responsiveness is relevant to considerations of the police-community connection. Many community-based groups that focus on issues of crime and disorder often do so in response to a particular threat, often embodied in a particular group.[11] What to do when teenagers are gathering in public space and engaging in voluble behavior that causes fear among some residents? What if there are suspicions amongst residents that some in that group are engaged in the sale of drugs or the trafficking of guns? How far should the police go in responding to these concerns? How do the police balance the various rights of the teenagers—the right to assembly, the right against intrusive searches and seizures—against their need to show that they take citizens' complaints seriously? Does a police force that takes civil liberties seriously threaten its own legitimacy?[12]

The drive to understand the police as importantly separate from society derives legitimacy not only from the need to protect civil liberties, but also from the police themselves, in their quest for professionalism. This is the second, and distinct, means by which separation can be legitimated. Indeed, the professional movement was initiated to increase the distance between the police and the communities they patrolled. In the machine-era of urban politics, officers were seen as too enmeshed in their neighborhoods, and thereby too easily corrupted. Professionalism was meant to make police more aloof, more controlled by legal and bureaucratic rules. Officers would thus be less corruptible. They would also be more efficient and effective in reducing crime.[13] This legacy of professionalism, and the separation from the community that it implies, lives on in police culture. Like other professions, the police seek to preserve a unique base of competence. From this base, they can resist community oversight.

The potential tension between subservience and separation derives, then, from both normative conflict and sociological practice; they can conflict as political ideals, they can conflict because of police adherence to ideals of professionalism. This is an intractable issue, because each ideal—of democratic decision making, of a neutral state that protects individual rights, of state actors who uphold professional norms—possesses considerable legitimacy. On those occasions when they do conflict, there is no ready calculus toward an ideal resolution. What complicates matters further is that there is yet a third way in which the state-society relation can be understood and legitimated.

Generative?

Rather than viewing the state as either subservient to or separate from society, one can also justifiably see it as fundamentally generative of society. In both democratic and liberal theory, there is an important sense in which state and society are distinct and need to remain so. But perhaps this helps us miss how the state is a basic creator of community. There are three ways in which this could be said to be true.

First, one could emphasize the extent to which state policy undergirds the conditions in which communities develop. Zoning policy, for example, critically shapes the class composition of a neighborhood, and thus helps construct a landscape of differentiated political capability. The provision of other social services, such as education, health care, and child care, also shape the well-being and political capacity of a community.[14] The formal social control apparatus is another contributing factor in a neighborhood's dynamic. One of the more potent critiques of the massive growth in incarceration in the United States is its impact on the male population of distressed neighborhoods. The absence of income earners and heads of families arguably increases those levels of distress.[15] In short, state policy structures urban dynamics significantly; the state helps generate the community that develops within a city's neighborhoods.

A second sense in which the state could be understood as generative of community emerges from discussions of governmentality. Working from an initial formulation by Foucault,[16] analysts of governmentality emphasize the means by which governance projects are constructed, rationalized, and implemented, particularly in liberal societies, where the political subject is presumed to be free. Projects of governmentality are thus, as Mitchell Dean puts it, "distinguished by trying to work through the freedom or capacities of the governed."[17] Analysts of governmentality place principal emphasis upon the rationalities of government: the expert knowledges upon which governmental programs draw and the implicit definitions of truth contained therein; the assumptions these knowledges make about the present and desired future state of the governed; the latent capacities within the governed that can be tapped in productive ways.[18]

Here our attention is drawn to how state agencies apprehend community. State actors employ certain grids of legibility upon the input they receive from the citizenry; they recognize some and not other forms of input as legitimate, they sort that input into categories, they react to it via certain prescribed routines.[19] James Scott provocatively refers to this broad phenomenon as "seeing like a state."[20] Community is thus not some independently existing entity, but is rendered sensible through a particular state epistemology.[21] The police are no different in this regard. As we will see, they construct community in particular ways that fundamentally impact the manner in which they respond to citizen concerns and demands.

There is yet a third way in which the state generates community. State efforts to improve legitimacy often involve trumpeting abstract ideals which state policy reinforces. These ideals are typically transcendent and moralized. Freedom, equality, justice, and opportunity are all obvious examples. Counterexamples are often mobilized as the target of state energies; terrorism is an evil because it restricts freedom, racial bias odious because it frustrates equal opportunity. In moralistically constructing state policy and its motivations, state actors create a more universal terrain that seeks to unite citizens into an overarching community. Various efforts to promote patriotism exemplify this. They work to create, in Anderson's words, an "imagined community" in whose name such dramatic action as war-making can be legitimated.[22]

Crime is one phenomenon often constructed in highly moralistic terms, in part to legitimate a robust formal social control apparatus.[23] As a moral plague, as an impediment to people's freedoms to possess property and to move safely through space, crime demands a strong state response. Certainly, the police understand their work in moralistic terms, as integral to a pitched struggle between the opposing forces of communal good and criminal evil.[24] By constructing this moral terrain, state actors like the police help to generate community. They seek to unite people around a common and transcendent vision of crime's genesis and suppression. Once placed upon this moral terrain, the citizenry are not primarily political subjects with rights and democratic potential, but instead potential victims who must be protected. These moralistic discourses work to limit the range of citizen discourses police officers consider legitimate; they are a set of ideological blinders with considerable consequences.

In each of these three ways—as creators of highly consequential state policy, as architects of an epistemology that defines legitimate forms of communal input, as protectors of transcendent goods—state actors generate community. These realities further complicate how best to understand, both normatively and sociologically, the relationship between state agencies like the police and the communities with whom they are meant to partner. The tension between subservience, separation, and generativity necessarily infuses the politics of community policing in ways that are not easily resolved.

The Police-Community Relation: An Empirical Approach

Just as we saw from debates about the political status of community, we see again normative disagreement, this time concerning the state-society relation. The state stands in no simple relation to its communities of constituents, and thus no single strategy for legitimation reveals itself unambiguously. The persistence of this ambiguity is even more obvious when one looks closely at actual police practice, including their encounters with the public. Such an examination reveals the presence of each of these modes of state-society connection. This renders even more illusory the simplistic narrative of a state eagerly responsive to citizen input.

There is ample reason to doubt that the police understand themselves, first and foremost, as subservient to the citizenry. Certainly, police-community forums in West Seattle are not instances of vibrant democracy. Instead, a predictable sequence of events occurs that limits open-ended discussion. Police officials arrive decked out in full uniform and place themselves at the front of the room. They are allocated a slot on the meeting's agenda, usually near the beginning. They outline the recent major crimes and explain their current or anticipated response. They tend to accentuate police successes; they trumpet the capture of any notorious alleged criminals. They often describe strategies for areas where long-standing problems are present, such as a home where suspected drug sales are occurring or a park that hosts anonymous sexual encounters.

After this monologue, the police solicit input. Sometimes audience members follow up on a situation the officer has already described, to ask a clarifying question or provide additional information. At other times, they ask questions about other phenomena of concern. Are there more abandoned cars on this block? Isn't graffiti becoming more of a problem on that street corner? In response, the police will explain their understandings of the pattern and indicate an interest in investigating further. Often, the officer will explain what limits their ability to investigate further. Perhaps there is disagreement with the Parks Department concerning how best to respond to anonymous sex in a park, perhaps the juvenile system does not stringently punish young thieves.

Discussion never emerged about the basic parameters of police action and authority. Rarely did citizens even attempt to so enlarge the discussion. The few attempts that did occur were ignored or deflected. The police's central mission and tactics were thus never open to debate; rather, they were a set of presumptions that necessarily limited the conversation. These forums were not instances where a true "partnership" was on display, where "co-production" was an ongoing reality.

Take, for example, a community council meeting at Blufftop, the public housing facility. Each of these monthly meetings included a presentation by the facility's police officer. Funded by the local housing authority, the officer focuses exclusively upon Blufftop. At this particular meeting, he wasted little time in proposing and propounding a new "no trespass" zone within the interior spaces of the sprawling complex. He argued that his encounters with people late at night led him to conclude that few were residents. Instead, he said, they were outsiders "without a legitimate purpose." His implication was obvious: they were there to traffic in drugs. The green spaces of the facility, in his words, were a "haven for unlawful activity." He wanted a means to police them effectively. The "no trespass" zone would allow him to cite any nonresidents found on the grounds during nighttime hours.

It was never clear just why the officer felt a need to get the council's approval for this measure; the nature of legal authority over these interior spaces seemed uncertain. It was also unclear whether residents could be cited for violating the proposed ordinance. The officer indicated that the law was "squishy" on this issue. Nonetheless, the residents quietly acquiesced to the plan. They asked only a few questions, none of which interrogated the plan's logic or consequences. When it finally came time for a vote, the president of the council tried to articulate the plan. After struggling to do so, she turned to the officer and said, "Whatever you want." The measure passed without dissent.

It may well be that the residents fully endorsed the officer's plan and agreed with his argument that loiterers were "broken windows" who needed to be removed. Indeed, interviews with some of these residents indicated that they were troubled by people they suspected of selling drugs. However, their lack of critical engagement with the officer's plan, their willingness to acquiesce to "whatever you want," was striking. To his credit, the officer sought community support for his plan, but what transpired was a cursory rubber stamp to a proposition that was wholly his own.

This consistent pattern of minimal discussion about police practices at community forums does not suggest that the police ignore public input. To a significant degree, police are responsive in this fashion; they recognize a degree of subservience to the citizenry they police. It is simply that this is not the only pathway to legitimacy that they develop. To understand how this is the case, I turn now to a review of the ways in which the various key projects of legitimacy—subservience, separation, and generativity—structure how the police understand and undertake their work.

The Project of Subservience

In any society where the democratic impulse possesses legitimacy,[25] state agencies must engage in practices that demonstrate some subservience to the public. This is especially critical in the case of the police, given their tools of coercion. The specter of a "police state" is ever present with armed agents of government authority. An appearance of subservience offers reassurance that coercive authority will not be exercised arbitrarily. This is a particularly resonant theme in American political culture, where wariness of an overly intrusive state has a long and potent heritage.

There are several means by which the police demonstrate their willingness to be responsive to the citizenry. Perhaps the most obvious example is the process by which officers are dispatched. Citizens can access the police department quickly, by dialing 911, and request assistance. That assistance arrives with uncertain speed, but genuine emergencies generate as rapid a response as possible. When requests are numerous, police work can be largely determined by them—an officer simply spends a shift responding to call after call. Here, the citizen assumes the role of customer, and the officer the service provider. The officer's work pattern is compelled by citizen demands.

Indeed, many critics of the police, and many senior police officers, bemoan the power of calls for service to dictate police practice. If an officer simply bounces from call to call, she is unable to spend extended time understanding and addressing an ongoing issue in a given neighborhood. Thus, the arguable need emerges to go "beyond 911"[26] to enable officers time to engage in often extensive "problem solving" operations.[27] Senior officers argue that younger officers lack precisely these skills. As one veteran said, "These officers go to a call, write up a report, and then forget about it. They think that's police work." He argued that the ascent of the dispatch system limited the development of a broader set of useful skills and orientations. He, and others, also argued that "aggressive dispatch"—the drive to get an officer to a call as quickly as possible—necessarily meant that officers were often sent to areas outside their beats. Officers thus spend less time inside their beats, and are less able to establish close relations with residents. The goal of "beat integrity," of keeping officers within the areas for which they are responsible, is thwarted. In these ways, "aggressive dispatch" may work against community policing, which emphasizes close ties with the citizenry and prolonged efforts at collective problem solving.

Regardless of the merit of these arguments, it is clear that the dispatch system represents the most obvious and significant way by which police departments are responsive to the citizenry. That departments endeavor to respond to calls quickly indicates a recognition that such responsiveness is critical to police legitimacy. Similarly, the fact that police officials attend regular community meetings indicates a desire to also demonstrate subservience to public input. As noted above, these encounters always include a time for public comment. That the ensuing conversation is often constrained does not eliminate the fact that the exchange occurs, and that the police feel compelled to outline a response to the information they receive. Officers seek to demonstrate that the information will be acted upon somehow, that a citizen comment deserves as specific a rejoinder as possible.

This willingness to solicit and potentially act upon citizen concerns is exemplified in the various reports the public can file with the police. These range from concerns about alleged criminal activity, such as suspected drug sales, to concerns about police practice itself, such as when officers use what citizens perceive as excessive force. The degree of police responsiveness to these various reports is subject to debate; as we will see, many residents wonder if such reports receive any attention at all. But they do represent a desire by the police to solicit public input and to demonstrate, whether sincerely or not, a willingness to respond.

Similarly, the police often work with multi-agency task forces that are created to confront particularly intractable problems. In Seattle, these task forces are referred to as "Neighborhood Action Teams," or NATS, and involve representatives from various municipal agencies. This wide participation should enable the government to respond in a coordinated fashion to issues related to public safety, such as land use, sanitation, lighting and other infrastructure, and ongoing criminality. The police participate deeply in these projects, and sometimes even spearhead them. In such a fashion, the police demonstrate an apparent willingness to respond meaningfully to public complaints about areas of longstanding concern.

The desire to respond to continued complaints about such longstanding problems motivates much of the work labeled "community policing." Community police team members in Seattle are not expected to respond to calls for service. Instead, community police officers are assigned to monitor those situations that generate continuous complaints from residents. Absent such long-term attention, in theory, these problems will resist amelioration. In short, the community police officer's role is understood as essential to the department's ability to be responsive to public concerns.

These are the most prominent ways by which the police pursue legitimacy through subservience. They solicit and respond to public input, they seek to alleviate concerns about crime and disorder. Through their everyday practices and their public interaction, officers implicitly recognize the citizenry's substantive role. Without such attempts at demonstrating subservience, the police's quest for legitimacy would surely suffer.

However, the police place limits upon this seeming subservience. They also understand themselves as significantly separate from the public. The drive for separation springs from two distinct sources. One comes from liberalism's insistence on state respect for individual rights, and from the constitutional and legal order that protects those rights. The other comes from the quest by state agencies like the police to develop and maintain professional codes of conduct.

The Project of Separation I: Liberalism and the Legal Order

As I emphasized above, the doctrine of liberalism emphasizes a politically neutral state. The state must not tie itself to any particular vision of the good life, but should build and defend a set of rights that enable citizens to pursue a wide range of individual expression. The expression of these rights is necessarily limited, to ensure that one person's development of her free will does not limit another's. But the state should not interfere beyond this point; it should not foist any proscriptions upon its citizenry beyond those necessary for society to function with minimal disruption to the right of expression and the general maintenance of order. Hence the importance of constitutions to liberal states, for these provide the basic architecture of what the state can and cannot do. State actions are thereby not dictated entirely by democratic sentiment, but also by constitutional and legal rules. These rules help divorce the state from the project of strict subservience. Otherwise, the liberal logic goes, the state might capitulate to a majoritarian program that could diminish the liberties and rights of marginalized minorities.[28]

Consider the example of community policing. On the one hand, police departments engaged in community policing are meant to solicit input from localized groups and to respond strongly to that input. In the name of localized democracy, state agents and active citizens are to come together and "co-produce" solutions to problems of crime and disorder. But what happens if a particular neighborhood group's concern involves the actions of another social group? What if, for instance, a group of largely Anglo homeowners is upset at their Latino neighbors, who perhaps pack several residents into a small apartment, or who play music at somewhat high volume? Here the police's responsiveness should be tempered by the legal code. If the Latinos are in legal violation of resident codes or of a noise ordinance, then action can be taken. Otherwise, the police should not pursue subservience by running roughshod over the law.

Indeed, the police do regularly cite the legal code as the basis upon which they act or do not act. Take the example of anonymous sex in a local park. The park in question contains a few acres of woods, with a trail system running throughout. It reputedly hosts encounters between gay males. These encounters generate continual complaints by residents of the adjoining neighborhood. These complaints, in turn, motivate the community police officer for the area to limit the sexual activity. However, he considers himself significantly constrained by the legal code, which sets fairly strict guidelines for when he can arrest someone. He needs to witness a sexual act in progress. Here he faces long odds: it is hard to move through the woods noiselessly, and he does not have the time or resources to engage in surreptitious surveillance. His tactics rely instead upon the occasional walk through the woods. He hopes that his presence will scare off those who seek liaisons.

I observed him using this strategy one afternoon. During a walk through the woods, he literally crossed paths with a middle-aged male. The man turned tail when he saw the police uniform. The officer caught up to the man, and asked questions about his presence in the park. The man appeared nervous. The officer expressed little interest in what the man did in private, but explained how he wished to keep the park free of illegal activities. Lacking evidence for an arrest, the officer dismissed the man after a brief sermon. However, the officer later suggested that the man's nervousness was a positive sign. If others would become similarly fearful, the officer hoped, then the illicit encounters might well decline. But this was a strategy that was clearly circumscribed by legal restrictions.

The legal code also emerged around an issue that energized many activists in a high-crime neighborhood in Centralia. These activists focused their attention less on those who engaged in such activities as drug delivery, and more on the landlords of the apartments where these people lived. The activists believed, as did the police, that the landlords did an inadequate job of screening their tenants and of regulating their behavior. However, the activists and the police found themselves limited in what they could do, because of legal regulations surrounding enforcement of landlord obligations. Many of these landlords were able to lease to tenants who possessed Section 8 vouchers (i.e., federally subsidized grants for housing). This meant that federal regulations helped determine actionable violations on the part of a landlord. In addition, the City of Seattle had its own regulations surrounding issues of building safety and other concerns, such as noise or accumulating trash.

The confusion surrounding these legal matters was evident in a small meeting in this neighborhood, attended by a handful of active residents and representatives from the mayor's office, the local housing authority, and the city's Department of Construction and Land Use. The meeting was called to inform residents of some possible new regulations the housing authority hoped to enact to enable them to punish landlords more aggressively. But the meeting served to generate as much confusion as hope. During a wide-ranging, nonlinear conversation, several matters remained unclear: how much legal power the housing authority or the DCLU possessed; what happened when federal codes conflicted with local laws; what legal authority activists had to demand an official investigation of a given property or landlord; and how one could mobilize the police as part of such an effort. In short, the welter of legal authorities and regulations limited what state actors could do in addressing the loudly voiced concern of activists focused on reducing crime in their neighborhood.

In this and manifold other ways, the law helps separate state actors and the community: the former can never entirely respond to the latter's concerns. In the case of the allegedly irresponsible landlords, the legal rules were designed to limit the intrusion of government officials into private property and to ensure a sufficient supply of affordable housing. This is but one example of the role of law in structuring the police's response to community requests. Even when a well-organized group of activists makes a persuasive case that landlords are culpable for their tenants' misdeeds, the police still cannot act forcefully. They are prevented, by law, from acting on simple subservience.

The Project of Separation II: Professionalism and the Expectation of Deference

The drive by state agencies, and especially the police, to separate themselves from the public stems from more than their legal authority. If liberal thought requires a state agency like the police to abide by legal regulations, and thus to be somewhat aloof from the vicissitudes of community pressure, it says very little about how such aloofness might be understood and legitimated by state actors themselves. Even minimal contact with police officers reveals the great extent to which they distinguish themselves from the wider public. They clearly see themselves as members of a unique group. Three principal factors fuel their construction of themselves as importantly separate from the citizenry: a desire for protection from community meddling; a felt need for undiminished authority in the situations they confront, in part to ensure their own safety; and a hope to attain status via comporting with standards of professional practice.

SEPARATION AS A MEANS OF PROTECTION   One afternoon, I was invited to share lunch with a group of officers in the precinct station's lounge. The lunch was prepared by a retired officer who enjoys coming to the precinct to cook a meal for the officers. The prospect of a warm, tasty, and free meal brings them together. I sat down to my plate of beef brisket and was introduced to the other officers at the table. One officer eyed me warily and promptly asked, "Are you a liberal or a conservative?" I had barely begun to answer when the officer launched into a monologue on the evils of the news media and its love affair with a small number of anti-police political activists. At issue, in particular, was media coverage of some then recent police shootings of African Americans. These shootings had generated much public criticism. The officer suggested that the media consistently discounted the sensibility of the shootings, most gallingly by not fully reporting the nature of the threat facing the officers involved. Other officers at the table murmured their assent, concurring with the broad argument that the media listens too closely to police critics and thus unfairly impugns the work of the police.

On another evening, I was asked to wait for a ride-along just inside the precinct station door, in an area where many officers mill about. One officer passed by a few times before stopping and only half-jokingly asking me, "Do you work for the mayor's office?" He said that he thought he saw me reading his name tag. He did not stick around to belabor the issue, but his point was clear: he was wary of the possibility that I could spy on him and use any information I might glean to punish him.

On yet another evening, I talked at length with an officer about efforts by some members of the Seattle City Council to conduct research to determine whether SPD officers engaged in "racial profiling." The council wanted to determine the racial composition of the citizens SPD officers stop for traffic violations, to determine whether a disproportionate number of those stopped are minorities. The officer was adamantly opposed to such a study, primarily because he distrusted the political motives of the council members. In his view, any such study would be rigged. As he put it, "If the City Council wants to find racial profiling, they will find it."

Each of these incidents illustrates a theme that was regularly sounded by the officers I encountered: that they are members of a politically vulnerable institution that can easily suffer the slings and arrows of ill-motivated activists and public officials. Officers express considerable fear that public meddling in police affairs will impact them most significantly in terms of punishments. This refrain was consistent—a plaintive wail that the police attract attention from uninformed whiners. These ostensive political operatives allegedly make baseless complaints about the police, thereby making the officers vulnerable to superiors who may sacrifice them to public pressure. Indeed, many officers assess their superiors largely on whether they protect the rank and file from such public pressure. Suspect superiors are castigated as too "political," too easily led to punish officers when the public complains. They want leaders to shield them from such pressure. From their perspective, the public is simply too ignorant to possess any power to determine their fate. The citizenry, many officers allege, can never understand the realities the police face and the necessary tactics they must employ. A police leader's job is thus to keep the public at bay, to insulate the officers from sanctions that might befall them should the public possess meaningful power.

In short, the police hope to be not only separate from the public, but actively protected from community intrusion into their affairs. Such intrusion, officers widely believe, can only bring negative consequences, including sanctions against officers whose split-second decisions will be scrutinized by ignorant and biased political activists. Thus, any talk by City Council members about racial profiling was seen as pandering to African-American activists who seek political benefits through ill-informed police bashing. If their superior officers did not protect them, they would be the fall guys for these machinations. As one officer ruefully observed, "Shit rolls downhill."

These dynamics fueled a set of political maneuvers by the police officers' union in Seattle. The officers were distressed that a new apparatus was established inside the organization to allow non-sworn civilians to investigate citizen complaints. The Office of Professional Accountability is the result of political efforts to provide some external oversight of police actions. The first high-profile case the OPA investigated involved an officer who detained a group of Asian Americans for jaywalking on a downtown street. Those detained alleged that the officer treated them rudely and that he singled them out because of their ethnicity. The OPA rejected the allegation of racial profiling in the incident, but did recommend that the officer be sanctioned for rudeness. The chief of police, less than a year into the job, accepted the OPA's recommendation. That led the rank and file to mobilize a vote expressing "no confidence" in the chief.[29] The message was clear: the officers expected the chief to protect them from such meddling by citizens, and publicly withdrew support from him when he appeared to sway too easily in the political winds.

This dramatic and well-publicized effort to highlight their alleged vulnerability to politics illustrates vividly a more pervasive sense in which the police understand themselves as embattled. Anything that smacks of "politics" is suspect, any effort in the name of civilian oversight represents an opening through which police opponents can seek the trophy of a punished officer.[30] Because even well-meaning citizens cannot understand the vagaries and challenges of the job, it hardly makes sense to the typical officer that anyone with an overt political agenda be given influence over police policy and conduct. Instead, the rank-and-file officer, charged with the task of responding quickly to situations where lives may hang in the balance, needs to be shielded from excessive community meddling. Separation from the public is thus understood as an important mechanism of protection.

SEPARATION AND AUTHORITY   Another means by which officers seek separation from the public is by expecting deference to their authority. When they arrive on a scene, the police expect those present to acquiesce to the officers' felt need to dictate the flow of action. They want citizens to obey orders and to comply with requests for information. The police believe they play a unique role in society, and they see their authority as critical to their success in that role. Citizens, in other words, need to recognize the police as a separate and powerful institution, and to defer accordingly.

In important respects, this desire for unquestioned authority is tied to officers' understandable preoccupation with safety. Given the unpredictable and potentially dangerous nature of some scenes to which they are summoned, officers sensibly hope to gain quick control amid chaos. Such successful dominion can thus reduce the possibility of danger. This functional connection between authority and safety is one I explore more in chapter 4. The important point to stress here is that a principal source of the separation the police construct between themselves and the citizenry is their desire for deference to their authority.

For example, in the story cited earlier of the man on the paths of the park known for illicit sexual encounters, the officer was quite distressed that the man initially sought to elude him. He berated the man for walking away, reminding him that it was "a good idea" to comply with a police request for a conversation. Another officer told me he derived great satisfaction from his job because, as he put it, "I like to win. I don't win them all, but I do win most of them." This officer is confident that, eventually, his authority will emerge triumphant, and he exults in that reality.

Indeed, the police regularly invoke their superior tactics and their capacity to exercise coercive force in describing their work. One officer described an encounter that occurred at a roll call attended by the mayor shortly before the World Trade Organization Ministerial Conference in 1999. According to the officer, the mayor attended roll call to urge restraint in the face of expected protestors. The officers responded derisively, and suggested to the mayor that he was underestimating the threat the protestors would pose to public order. The mayor reportedly bristled at these remarks, a reality that the officer dismissed. He said, "We have guns. We aren't bashful. Don't come into our station and expect us to shut up."

Another sergeant evinced much the same logic in concluding a dissection of an encounter I had just witnessed, wherein he negotiated a standoff between a shop owner and a towing company. The shop owner had parked his car in front of a fire hydrant near his shop, a violation that can be punished with an immediate towing. A traffic officer had ticketed the car, and summoned the tow company with which the city contracts. When the tow truck arrived, the shop owner hurried out to his car and occupied it. The truck operators stopped the tow and called the police because they would not tow the car with the owner inside. When they arrived, the police were told by the shop owner that he suffered from a heart condition, and thus any effort to remove him forcefully might cause a heart attack. For their part, the tow operators were unwilling to unhook the car without receiving payment from the car owner. The sergeant artfully negotiated a compromise, whereby the shop owner agreed to pay a reduced fee in exchange for the cessation of the tow. The sergeant explained his reasoning for seeking an amicable solution: "We try to avoid a fight. But if they really want to fight, we'll fight."[31]

At the limit, then, the police recognize their need to prevail, and will physically enforce that authority if necessary. Their understandable preference is for their authority to be immediately recognized and deferred to. Absent such deference, officers will assert their authority as they see fit. They recognize themselves as distinct and separate from the citizenry, possessors of a unique source of power that will be hegemonic.

SEPARATION AND PROFESSIONALISM   Beyond their adherence to the law, then, the police reinforce a sense of distinction between themselves and the public by seeking protection from excessive political intrusion into their affairs and by embracing a self-construction as authoritative actors. An additional source of separation comes through the prestige associated with the status of a professional. Recall that the ideal for police organizations that preceded community policing, the professional model, stressed the unique bases of expertise the police could mobilize to reduce crime. The police sought to make themselves analogous to other high-status occupations. Just like doctors and lawyers are specially educated, so were police officers trained according to high standards that ensured an efficient workplace. This was tied to the effort to wage an effective war against crime. Because of their superior tactics, the police would consistently outwit criminals, and therefore deter future criminality.

Even if one could argue that community policing has supplanted the professional model,[32] the notion of the police as possessors of unique expertise clearly persists, and for officers this expertise reinforces a distinction between themselves and the public. For example, during a ride-along, a dispatcher asked an officer to respond to a citizen complaint about a group of men who were selling drugs on a street corner. The officer dismissed the call because the citizen "had not been trained" to detect drug sales. It would therefore likely be a waste of his time to respond. Another officer I accompanied observed a group of young men dispersing from a street corner as his cruiser drove past. This was a significant phenomenon, he asserted, one that only a "trained police officer" would notice. Their dispersal, he argues, is a clear signal that they are "up to no good." Why else, he wonders, would they be dispersing?

This notion of professional separation was reinforced subtly but powerfully at police-community forums. Each conversation between officers and members of the public focused on some matter related to crime and disorder. In each case, the officers controlled the conversation, reserved for themselves the last word, and discussed—often at length—the various measures they would take to address the issue at hand. If there were any obstacles to the police's effectiveness, these were invariably matters beyond police control: a too-lax system of punishment, an unresponsive bureaucracy, the delicate politics of arrests of homosexuals. In short, the police reinforced a notion of themselves as the expert professionals, able to deploy their unique set of knowledge, skills, and tactics to solve a problem beyond the scope of an ordinary citizen to confront.

In sum, then, the police construct themselves as a separate and powerful social group, and legitimate this distinction in potent and reinforcing ways: they are legal actors with allegiance to the legal code; they are members of a politically vulnerable group that deserves protection from ill-informed public meddling; they possess an authority to control situations to which the public should defer; they command a unique base of knowledge, and thus deserve an elevated professional status. From this position of separation from the public, it is obviously difficult for police to approach the citizenry as fully vested "partners" with whom they can "co-produce" strategies for crime reduction. In this way, subservience is limited by separation.

The Project of Generativity

Subservience and separation do not exhaust the means by which the connection between the citizenry and the police is constructed. In each of these approaches, state and society are considered distinct, and the state is either responsive to or shielded from society. However, the two approaches neglect the significant ways in which society is very much a product of the state, how the two are not neatly distinguishable. As I elaborated above, there are three principal means by which the state generates community: by constructing policies that significantly determine the material realities within given neighborhoods; by establishing a set of routines and practices by which community input is understood and processed; and by constructing a moral universe that state and community together inhabit and seek to protect. Through these policies, practices, and discursive constructions, the line between state and society is blurred in ways that are critical to state efforts at legitimacy.

It is a point of some controversy whether the police do much to shape the material realities that exist in given neighborhoods. There is evidence that the police can help reduce the fear of crime, but the evidence that they can actually reduce crime is largely missing.[33] In terms of the structural dynamics that shape neighborhoods—the type and caliber of the housing stock, the localized labor market, the provision of social services—the police possess no impact whatsoever. So, the police's capacity to generate community primarily involves the epistemology they mobilize and the moralistic understandings they develop.

THE EPISTEMOLOGY OF COMMUNITY The rhetoric of community policing suggests that the police can and should work closely with the "community." But just what does this term mean to the police? What do officers recognize as a community? Who do they consider capable of speaking for a community? In general terms, how do the police channel and make sense of the input they receive from the public?[34]

What was striking from my conversations with police officers was the fact that the term "community" possessed very little resonance. I regularly asked officers about the communities in the areas for which they were responsible. Were there identifiable such communities? Where were they? Of whom did they consist? Were they aware of any formally or informally organized community groups focused upon issues of crime and disorder? If so, how did they understand the connection between any such groups and the police?

These conversations revealed that the officers' geographies were structured primarily by those "problem locations" to which they were summoned most regularly. The key locales on their mental maps were the hot spots that elicited regular calls for service: a public housing facility; a block where outdoor drug sales were apparently common; a row of adjacent houses reputed to be home to gang members. In some cases, officers knew well many of the residents in these areas of particular concern. One officer, for example, took me down an especially notorious block and briefly described the residents of nearly every single building.

But that same officer knew almost nothing about areas to which he was not regularly summoned. As we drove along a major thoroughfare, he waved his hand at a neighborhood to our right and said, "That's all residential." That area, he implied, was full of stable homeowners, and thus of no interest. Indeed, nearly every officer with whom I rode spent their discretionary time either cruising along major streets or visiting hot spots. Neighborhoods that did not host ongoing problems were largely invisible.

The term "community" was not one that officers used regularly in describing their areas. It was not a term that emerged organically from our conversations about the geography of their beats. It was the extremely rare officer who possessed information about the informal dynamics in a given place, who could provide keen observations about the realities in any particular neighborhood.[35] Any such detailed knowledge was almost exclusively confined to community police team officers, although even many of those struggled to provide comprehensive explanations of ongoing problems. For their part, officers in regular patrol might know quite a lot about a particular problem location or two. Yet their knowledge was limited to the alleged perpetrators of those problems, and did not include the wider dynamics in the neighborhood. As I explore in detail in chapter 4, problems were largely considered the handiwork of a few "bad apples" who plagued an otherwise peaceable area. The term "community" was rarely discussed as either a constituent component of these problems or as a significant player in their resolution. As one patrol officer admitted, "I don't think too many of us spend much time thinking about community."

Even if "community" was not central to officers' understandings of place, still the police were regularly summoned by citizens who sought assistance with matters of concern. Invariably, these requests for assistance are filtered through a set of bureaucratic and cultural scripts; public comment is translated into a language that the police recognize. So community input does not exist independently of the police, but is sifted through officers' screening mechanisms.[36] The police actively construct community through these translation practices.

For example, an officer was summoned to a residence because of a "911 hangup" call. Someone in the house in question had called the emergency number, and then quickly hung up. Such a call becomes an emergency, for fear that the caller's request for help has been interrupted by a dangerous assailant. The officer hurried to the residence in question, and arrived simultaneously with his superior, a sergeant. A man who lives in the house with his wife was outside, repairing his car. He explained to the officers that their car was vandalized, and that his wife called the police department. The sergeant was visibly angry, and explained why a 911 hangup is an emergency call. Further, he explained how he and his officer were pulled away from other pressing matters to respond to this call. Dissatisfied with the man's explanation, the sergeant called the dispatcher to gather more specifics. The dispatcher could not provide any more information, but then the man's wife emerged from the house. She indicated that she did indeed dial 911, but hung up when she realized that there was a non-emergency number to call instead. Both she and her husband were profusely apologetic, although the officers remained visibly bothered throughout the conversation.

In this instance, we see quite clearly how the police's apprehension of public input is structured by various bureaucratic routines. What to the couple in question seems like an ordinary act—to hang up the phone when one realizes one has dialed a wrong number—becomes a matter of concern and frustration for the police. The sergeant, in particular, struggles to accept the fact that the couple failed to appreciate the implications of the woman's actions for the police. He works hard to understand precisely what happened, and explains, with detail and exasperation, how their actions were interpreted by the police. He obviously resents this needless interruption to his day, and he does little to disguise his feelings.[37]

In another case, I accompanied an officer who was part of a group responding to a call involving a domestic disturbance. A long sequence of events culminated in a recommendation that the man involved move his belongings out of his girlfriend's apartment. As the group waited for this task to be completed, a call came over the radio concerning a report that some shots were fired in a nearby neighborhood. The officer turned to me and explained that this was a "nothing" call. She indicated that unless there are several such reports, there is no sense in responding. Only then, she said, would the police know that this was a call of substance and thus worthy of response. I could not help but wonder about the resident who had phoned the police: did he or she believe this was a "nothing" call?

On yet another evening, an officer was summoned to an apartment to file a report of a stolen car. This would involve much paperwork, but he did not wish to do all of this at the residence in question. Instead, he pulled his car into a parking lot in a neighborhood that generates many calls for service, some of them involving violent crime. The officer was anxious to respond to any such calls. He reasoned that he did not want to be in an apartment doing paperwork when a priority call arose. He would rather minimize his time in the apartment to maximize his capacity to respond to an emergency. After several minutes spent doing paperwork in his car, he ultimately arrived at the apartment. There, we discovered that the owner of the stolen car was waiting for the police so that she could get to work. Further, she needed to arrange care for her small child. The officer's tardy arrival thus reduced her work time and complicated her life.

Each of these instances illustrates a broader point: that "community" and "community input" do not exist independently of the police's construction of them. In each case, the police translated a request for service into a particular form, one different from the form generated by the citizen. An innocent hang-up becomes an emergency, a frightening report about a weapon becomes a nothing call, a request for assistance on a stolen car becomes a restriction on an officer's ability to respond to something more interesting.

These sorts of disjunctures produce much tension between citizens and the police. What I wish to stress now is that the community and the police are not strictly separable entities. Instead, the police construct community through their various routines and epistemologies. In such fashion, the state generates community; it defines what, in practice, community actually means. Something similar occurs when the state constructs a moral understanding of its actions.

MORALITY AND THE POLICE-COMMUNITY RELATION   One of the most potent and consequential ways by which state actors construct community is by developing a set of moral justifications for their actions. Recall the lieutenant's performance discussed at the outset of this chapter. What made his rhetorical performance so powerful was the sense of moral indignation that undergirded it, the palpable outrage at a punishment apparatus unable to address the threat posed by unfettered car thieves. Cast onto this terrain, car thefts are far more than an inconvenience—they are a plague inflicted upon vulnerable citizens. Further, these citizens deserve more protection than they get. It is the state's moral obligation to do more. By casting criminality in such moralistic terms, the lieutenant fuses state and society in a particularly powerful way. No longer separate entities, each with some influence upon the other, state and society are situated together in a moral combat zone, a transcendental realm where good and evil battle. "Community" thus emerges not as a constellation of citizens with sufficient political agency to help oversee police policy, but rather as a site upon which the virtuous can defeat the unsavory. The police and the citizenry are implicitly joined in an unquestionable pursuit of a moral victory. Note how the lieutenant says, "I'm going to tell you the truth, and it's going to irritate everyone in this room." The possibility of dissent over the moral terms of the battle is inconceivable; state and society are joined in an overarching struggle to ensure that the good emerges victorious.

This sense of moralism is entrenched within the police's subculture. The coercive authority the police possess is often justified in terms of the abiding good that police work enables and protects. This narrative cloaks police authority in a penumbra of exalted justification; it legitimates coercion by emphasizing the greater good that coercion protects. One officer, for example, engaged in a rough search of an African-American man he thought might be connected to a robbery in a home occupied by a ten-year-old boy. After the man complained about the search, the officer legitimated his actions by emphasizing his need to respond to the mistreatment of a vulnerable child. He sought to absolve himself of guilt for his forceful behavior by connecting to the higher purpose of protecting the innocent.

This self-image, as protectors of the weak and vulnerable, is one that officers project frequently. One officer allowed a conversation about citizen oversight of the police to become a vehicle for his exasperation. He recalled an incident he had handled a few days earlier, where a child was under threat from his mother's boyfriend. When the officer arrived at the mother's home, he was astonished to find that she had barricaded herself into a room alone, leaving her child vulnerable. The officer noted that it was left to him to fulfill this basic responsibility of protecting her child. He drew a clear conclusion: if members of the public were unwilling to place themselves on the line for their own children, and wanted the police to do so instead, then they were in no position to criticize the police's work.

This exalted status for police and their crime-fighting work also makes it hard for officers to understand how they might appear to citizens as anything other than beneficent. One officer was astonished that a citizen complained to his captain because the officer stopped by for a "knock and talk." This is a practice whereby officers knock on the door of someone they think might be connected with a crime or might possess information about criminal activity. The officers use it to ferret out information, presumably with the compliance of the person involved. In his own view, the exasperated officer merely wanted to "have a conversation." Why, he wondered, would any complain about that? He could not understand why an officer in uniform symbolically represents something more than a casual conversant, why a citizen would possess anything other than complete faith in the police's mission.

A central part of this mission, for the police, is the exposure to danger that officers necessarily embrace. They are acutely aware of the risk they assume by inserting themselves into scenes of potential danger and violence. Even if many of them eagerly embrace this challenge—many are quick to respond to high-priority calls, with lights and sirens blazing, even from several miles away—they also exalt the potential sacrifice involved. It is of telling significance that officers who died in the line of duty are featured on wall displays in police precincts, a perpetual reminder of the life-threatening nature of police work. It is perhaps not surprising to hear a police officer refer to his colleagues as "men of honor." He knows that they are pledged to sacrifice themselves to defend the citizenry they are sworn to protect. Such sacrifices are made worthwhile by the moralistic language officers use to cloak their work; lives lost are legitimated because they preserve the greater good.

To situate their work firmly within the transcendental realm of the good is thus an understandable cultural response by officers to the dangers inherent in it. But such moralizing has implications for police-community relations. To so exalt their work is to thereby diminish the legitimacy of public oversight. If their work supports the greater good, then it lies beyond question. Recall again the lieutenant at the neighborhood council meeting. His outrage over the lax treatment of car thieves was implicitly a moral message about the danger of crime and the need to respond to it forcefully. The potency of this morality helped make it impossible for him to imagine a serious alternative to his framing of the situation. An alternate perspective on police work, perhaps one mobilized by critics, thus possesses little or no legitimacy and can be ignored or actively disparaged. Broad, open-ended discussions with the public about the direction of police policy and practice are hard to countenance.

In short, the separation between state and society is regularly blurred, in ways that restrict the reach of citizen oversight. State agencies like the police actively construct and situate the citizenry. Public input is filtered through various ordering schemes, community members are placed on a moralistic battleground where good faces evil. Community is thus not something separable from the state, but is generated by state actors like police officers. Through their definitional and moralistic work, the police construct community in particular ways, with telling impact on the nature of state-society relations.

Conclusion

A simple, uncomplicated, perhaps nostalgic vision of state-society relations infuses much of the rhetoric of community policing. According to this ideal, residents of urban neighborhoods can come together and develop a unified vision and resolution of their problems. Further, the police are to approach these neighborhoods in a spirit of responsiveness and cooperation. Officers should embrace neighborhood understandings of their concerns and should work as co-equal partners in crafting and executing strategies to address those concerns. The state should be subservient to a citizenry that exercises its political agency with alacrity and cohesion.

In some ways, the police respond to this imperative to subservience: they provide a range of services to meet the demands placed upon them. They respond to calls for service, they attend public meetings, they solicit citizen complaints and concerns via various mechanisms. But subservience by itself hardly characterizes the police-community relation; separation and generativity also figure large in the state-society connection. As much as the police demonstrate responsiveness, they also persistently reinforce a self-image as importantly separate from the citizenry. Such separation is often legitimated in terms of the police's need to enforce an abstract legal code that helps protect individual rights, and rightly so—the police do need, at times, to resist unjust demands from parochial social groups. But the drive for separation is fueled by other factors as well. Officers seek protection from their critics, they defend their authority to control the scenes to which they are summoned, they exalt their professional status vis-à-vis uninformed citizens.

Similarly, they exalt their work in terms of an overarching moralism that joins state and society together in an imagined community of virtue, wherein the evil stains of criminal pollution deserve an unquestioned vanquishing. In this way, and in a set of translation practices through which community input is rendered sensible, the police construct community in their everyday practices. Community is thus not something distinct from the police, but the result of a set of cultural understandings and bureaucratic routines.

In other words, the state is not simply a black box, a coherent object of abstract political theory. It is very much a social creation, developed through a complicated set of practices and cultural scripts. This reality complicates the police's pursuit of legitimacy. Whatever the resonance of the narrative of state subservience to legitimations of community policing, it is not the sole means of understanding the state-society relation. As the police demonstrate daily, any tendency toward subservience is balanced, and at times subverted, by the projects of separation and generativity. As a consequence, "community" stands in no simple relation to the police. A community may, at times, possess some capacity to mobilize a police response. But its members are frequently dismissed as either misinformed or the passive recipients of the police's expert and virtuous efforts.

The processes of separation and generativity thus limit the extent to which the "community" possesses meaningful oversight of the police. This becomes obvious when we open up the black box of the state and witness the competing imperatives toward police legitimacy. Another advantage of opening this black box is that it provides analytic purchase on the cultural dynamics of the police. Because these dynamics are a critical component of the police's stance toward the citizenry, particularly as part of community policing, they are the focus of chapter 4.

"Don't Drink the Kool-Aid": On the Resistance to Community Policing

In the 1990s, the Seattle Police Department attempted an ambitious reform effort. Under the leadership of Chief Norm Stamper, who had been brought to Seattle from San Diego, the department tried to reorient all of its operations around the philosophy of community policing. For Stamper, this meant making the police an agency through which myriad efforts at neighborhood betterment could be channeled. The police would listen closely to a wide range of citizen complaints, and would address as many of them as possible. This might mean soliciting assistance from other branches of the state's bureaucracy on matters outside the police's expertise. But the mandate of the police was deliberately broad: it was to be the linchpin agency in neighborhood efforts in self-improvement.

Stamper recognized that such a thoroughgoing effort at organizational change would require the support of the rank and file. Given the amount of discretionary authority low-level police operatives possess, and given the well-documented pattern of effective resistance to similar efforts at police reform,[1] Stamper's commitment to devoting attention to ensuring "buy-in" from lower levels of the hierarchy was clearly well-advised. To this end, he regularly invited officers to short seminars on the philosophy of community policing. The goal was clear: to encourage officers to accept community policing as their principal vocational orientation.

These seminars remain the butt of frequent sarcastic jokes inside the SPD. Commonly, officers refer to invitations to the seminars as requests to "drink the Kool-Aid." The reference here is to the 1978 Jonestown Massacre, when more than 900 followers of Jim Jones drank cyanide-laced Kool-Aid rather than surrendering their territory in Guyana. Some officers ridicule the language used at the seminars, in which leaders allegedly discussed "the rivers of community policing" to emphasize the breadth of the reform movement. The persistence of such sarcasm illustrates the hostility with which many view the project of community policing.

Why such hostility? Why would officers refuse to re-orient themselves as Stamper advocated? Why resist efforts to deepen the police's connection to neighborhoods, to broaden their mandate in ameliorating the problems citizens enumerate? Why is organizational reform akin to committing suicide?

In chapter 3, I showed that police impulses toward subservience to the citizenry are tempered by the projects of separation and generativity. The importance of these latter two projects limits the police's ability to envision citizens as co-equal partners. This reality has telling implications for the success of community policing. Why is the calibration of the police's mission in this fashion so difficult for officers to accept?

My arguments here extend those of chapter 3. I open up further the "black box" of the state and outline the internal dynamics most critical to police resistance to community policing. In the first section, I explore the importance of the capacity to exercise coercive force to particular tendencies within the police subculture, tendencies that reinforce the police's sense of themselves as members of an authoritative agency sharply distinct from the citizenry. I follow this by exploring how this authority gets expressed organizationally. I focus particularly on how different actors within the bureaucracy work in relative isolation to attain an independent measure of prestige. These emphases on specialization and discretion lessen the organization's capacity to coordinate officer efforts to solve problems identified by the public. I then explore the common narrative officers employ to understand criminality. This hegemonic story describes crime as acts of a few "bad apples" within a given area. By extension, the project of policing is to excise these bad actors through arrest and incarceration. Importantly, such a narrative implicitly reinforces the power of the police as the agent of expulsion. Citizens are thereby viewed largely in terms of their ability to provide information about those who need to be expelled. The public is not understood as the police's co-equal partner.

Collectively, these internal dynamics help thwart the project of sub-servience in everyday practice; they prevent more equitable partner-ships between the police and neighborhood groups. Officers' collective sense of themselves as capable, authoritative actors extracting evildoers limits the extent to which they can embrace community policing. By encouraging each other to resist change, the officers reinforce their sense of separation from the public. They thereby obstruct the demo-cratic thrust central to the legitimation of community policing.[2]

This sense of separation frustrates many citizens, as is explored fur-ther in chapter 5. For now, my goal is to explain how dynamics within the police organization motivate the effort to ensure that officers do not "drink the Kool-Aid."

Coercive Force and the Cultural Fundamentals of the Police

Any explanation for why the police tend to keep the public at arm's length must take seriously several components of police culture and organization. Like all social organizations, the police never enact them-selves in a simple, uniform fashion. It is far too limiting to consider the police solely as one component of some abstract entity called the state whose direction can be determined or directed by normative political theory. Instead, we need to understand the internal dynamics of the police—the key components of their culture, and the central narratives they use to understand themselves, their work, and their relations with the public.

Of course, these internal dynamics are critically shaped by external forces. The most significant of these is the designation of the police as the state's primary agent of coercive force. As Bittner long ago recog-nized, what unites the disparate actions police are asked to undertake is their capacity to compel a response, forcibly if necessary.[3] Police are summoned to such a wide array of scenes because someone believed that coercive force might be necessary to resolve a problematic situation.

One critical implication of the police's capacity to exercise force is its influence on the subcultural world officers construct. Dominant features of that culture can be traced to the coercive role officers are equipped (literally) to play.[4] I focus here on three main components of that cul-ture.[5] One is an emphasis on a masculinist notion of adventurousness embraced by many officers. A second is an understandable preoccupa-tion with safety. The third—and here I revisit some of my discussion in chapter 3—is the widespread belief that police authority needs to be asserted effectively. Each of these subcultural components can be traced back to the coercive authority of the police. And each underlies officer resistance to a robust role for the citizenry in shaping police practice.

Adventure/Machismo

As the state's principal agency of coercion, the police are regularly summoned to potentially dangerous situations. Citizens call the police when under physical threat, and ask officers to deal with those who pose such a threat. The police are thereby obligated to insert themselves into situations where the possibility of violence looms, and to use violence themselves if necessary to protect the vulnerable. As Bouza describes the transformation that accompanied his initial entry into the police fraternity: "Once I put on the badge and uniform, I was expected to run toward things I'd formerly run away from."[6]

Many officers embrace this challenge. Such officers quickly volunteer to handle calls that represent risk; they seek opportunities to turn on lights and siren in pursuit of a particularly notorious perpetrator. They use discretionary patrol time to unearth chances to exert their authority; they cruise in higher-crime areas hoping to encounter foes of the police against whom they can test their mettle. The valorization of this approach to policing means that officers assess one another in terms of their willingness to handle dangerous calls. Those who seem reluctant to do so are disparaged as something less than full-fledged police officers. Some police, for example, joke about older officers with ebbing energy for proactive work as being "retired on duty." Another officer, after a conversation with a colleague who wished not to pursue a suspect who had fled a scene, derided him by saying, "Obviously, he has a motivation problem."[7]

This cultural celebration of adventurousness generates two important implications for community policing. One is that many officers ridicule the slow-paced, heavily interactive work of engaging the citizenry. There is little adventure in community meetings, during which the conversation predictably drags through minutiae and includes performances of pettifogging. Nor is there much excitement in trying to coordinate with other city agencies to improve poor lighting or prevent the dumping of unwanted material in an alleyway. Indeed, such aspects of community policing are commonly dismissed as "social work," and are implicitly feminized as such. These are not matters for the strong, masculinist, adventuresome officer. Further, officers who work patrol rarely stoop to associate themselves with those involved in community policing efforts.

Rather, community policing is regularly denigrated as a refuge for those who seek to avoid hard work and danger, a place to "work banker's hours" and avoid the rough and tumble that "real police work" necessarily entails. I explore this bureaucratic isolation of community policing in more detail later in this chapter. For now, I note that an important contributor to that isolation is the association between community policing and the more feminized work of relationship building.[8]

This is not lost on community police team members themselves. On several of my ride-alongs with members of this team, I witnessed officers jumping at opportunities to respond to radio broadcasts concerning potentially violent situations. In some cases, the officers turned on their lights and sirens to respond to calls from locations that were several miles away. Because the intervening streets were heavily trafficked, their response time was likely to be slow, a factor that called into question their willingness to respond. Similarly, an officer serving as the community policing officer for Blufftop rejected a suggestion by the project's manager that he schedule regular hours in a publicly accessible office so that residents could approach him with their concerns. The officer indicated that he would not waste his time "sitting in an office and playing solitaire." Instead, he patrolled the streets in search of street-level drug dealers. The lure of a major arrest trumped the boring work of community interaction.

This desire to remain connected to the adventuresome side of policing helps explain enthusiasm for a short-lived "power community policing" model with which their unit experimented during the period of the research. Instead of leaving the officers to their own devices to address the problems in their areas, the "power" model coalesced the several members of the unit. As a group, they could engage in concentrated anti-crime activity in areas of ongoing concern. On occasion, for example, officers created a "buy-bust" operation to arrest street-level drug dealers. The community police officers enjoyed this taste of proactive, adventurous policing and were displeased when the experiment was terminated. So, even for those officers ostensibly committed to the slow, patient work of community building, the call of adventure continues to beckon.[9]

A second important implication of the embrace of adventurousness is its tendency to legitimate brusqueness. If officers seek to encounter and overcome the more dangerous members of the citizenry, they recognize the need to establish their greater power. Unfortunately, tools for isolating those in need of such displays of power are crude. As a consequence, the police sometimes assert their superiority against those who present no appreciable threat. And when officers, in the words of one resident, "act all macho," it can erode public trust in the police. The possibility that adventurousness will erode the tenor of police-citizen interactions also holds true for officers' embrace of safety.

Safety

The centrality of safety to the cultural world of the police is understandable. Given that the police's coercive capacity compels them to enter dangerous situations, it is hardly a surprise that officers do a great deal to protect themselves. Safety is reinforced in a host of ways, from formal training in tactics to gentle admonitions to one another to "stay safe out there." Officers are encouraged to make safety a paramount concern and to approach situations carefully to maximize their own protection.

One important manifestation of this is officers' treatment of suspects, particularly those officers believe might pose some threat. Officers are likely to engage such suspects verbally only after first ensuring that they are not dangerous. The typical means for doing so is a "pat down" of the citizen in question. Because officers are legally permitted to conduct such "pat downs" as long as they can demonstrate concern for their safety, they regularly employ them.[10] Unfortunately, such suspicion is not always directed accurately.

To illustrate this point, let me return to a scenario I discussed briefly in chapter 3. An officer I observed spent about an hour one morning trying to track down an African-American man, Ethan, who had allegedly barged into the apartment of his girlfriend and taken $500. According to the woman's son, the suspect had knocked on the door of the apartment, brushed past the boy when the door was opened, found $500 in the woman's bedroom, and dashed away. The officer got a statement from the boy and began scouring the neighborhood for information about Ethan's whereabouts. He visited several apartments and gathered a rough picture of the suspect's physical characteristics and clothing. At one point in the process, he looked up the block and noticed an African-American man walking down the sidewalk. The officer quickly jumped into his car, surged up the block, parked, jumped out and commanded the man to place his hands on the hood of the car. The officer conducted a hurried pat down, brushing off the man's questions. The officer said, "I'll explain everything in a moment," then proceeded with some questions about Ethan. After gaining little insight from the man, the officer briefly explained that he was looking for Ethan because he "assaulted a ten-year-old kid." Unsatisfied by the explanation, the recipient of the search asked the officer, "Do you do this to everybody?" The unstated message was clear: the man believed he received such rough attention because of his skin color.

For his part, the officer later justified the search because he thought the man matched the description of Ethan and thus represented a potentially dangerous suspect who might elude capture. The officer believed his safety was sufficiently at risk to legitimate the pat-down. Regardless of this logic, the result in this case appeared to be a diminution of police legitimacy in the eyes of at least one citizen. When this scenario is repeated in Seattle and other cities, even in the name of the much-vaunted goal of safety, it can compromise the closer police-community connections community policing promotes.

Concerns for safety also lead some officers to resist efforts to reduce the physical distance between themselves and citizens. These include, most prominently, efforts to get officers to patrol via some mechanism other than a patrol car. Because such alternate modes of transport—foot, bicycle, horse—create less of a barrier between officer and citizen and thus ostensibly promote a stronger flow of communication, they are frequently touted as important facets of community policing. But the patrol car is tied to officers' sense of safety. It enables officers to enter and exit a scene quickly, which might be important tactically. It can provide an actual shield between officers and potential assailants. And it contains an in-car terminal that allows ready access to a database from which officers can retrieve information about a suspect and a situation. This helps officers to assess the dangerousness of a scene they are about to enter. Even community police officers, whose mission revolves around citizen contact, express reluctance to eschew the patrol car, for fear of compromising their safety.

Concerns for safety also help to explain why officers are preoccupied with ensuring that their authority is unquestioned in their interactions with the citizenry.

Authority

The potential dangerousness of police work, then, underlies both the masculinist embrace of adventure and the pervasive concern for safety. It also influences the cultural emphasis on police authority. To exercise their coercive power legitimately, police officers need to understand quickly the nature of the scenes to which they are summoned and to determine whether and what kind of threat exists. This is especially important in chaotic situations, because danger may lurk. It is thus unsurprising that police officers place tremendous importance on their capacity to assume authority. This is obviously connected to safety, and is witnessed most clearly by the tactical training officers receive on how to approach and stabilize scenes of potential danger. The best way to avoid force, and to use it wisely if necessary, is to approach a scene properly and to quickly dictate the flow of action.

This felt need to assert authority becomes ingrained.[11] Even in situations with minimal danger, officers often conduct pat-down searches or request that citizens position themselves in particular ways. For example, one officer I observed demanded that a teenager with whom he was conversing stand up. This made it easier for the officer to look the young man in the eye during an interrogation about goings-on in a nearby park. In another instance, a sergeant decided to act authoritatively after a long interaction with a young man during an on-street investigation of an apparent gang-related shooting. The young man was one of three passengers in a car that matched a description of one near the scene of the shooting. After establishing that the young man had been less than truthful, the sergeant pointed his finger toward the young man's face and declared, "This is what happens to people who don't tell the truth to the police." The sergeant handcuffed the young man and escorted him to the back seat of a patrol car.

In these and many other instances, officers seek to establish their authority, to dictate how a situation will unfold. This capacity to establish authority is obviously functional to the core police task of exercising force, and can be of considerable public benefit in many situations. In cases of domestic violence, for instance, the safety of all involved is contingent on the rapid establishment of order amid chaos. Of course, overly brusque assertions of authority can exacerbate chaos. On a call involving an obviously drunken woman who physically resisted a sergeant's effort to prevent her from entering her home, some fifteen officers responded, and sought to handcuff the woman. The fervent activity and loud admonishments only served to fuel the woman's resistance. Finally, a female officer stepped forward, talked calmly to the woman, and convinced her to relax sufficiently to allow the arrest.

So, even in situations of some disorder, crude assertions of authority can produce more chaos, not less. In terms of making the police more consciously subservient to the public, the cultural assumption of police authority poses obvious problems. If this authority is simply presumed, then police officials, by definition, do not consider the public as co-equal partners. The challenge for the police is to distinguish between emergencies where their authority is necessary and situations like community forums where open discussion can be welcomed. What is a defensible cultural response to potentially violent situations becomes a liability when the police are legitimately asked to open their practices to debate. To an extent, one can empathize with the patrol officer who lamented to me that the public does not fully comprehend how he "can't be Officer Friendly" when he believes his first priority is to assert his authority. But this cultural response to the street-level realities the officer faces can reduce the space for the political efficacy of the citizenry.

In sum, the police's core coercive function is translated into cultural emphases that drive the imperative to not "drink the Kool-Aid." As officers reinforce masculinist notions of adventurousness, of safety, and of deference to their authority, they implicitly reduce the political agency of the citizenry. To build more truly co-equal partnerships involves a feminization of police work and erodes the police's self-construction as authoritative actors who ensure their safety by commandeering the scenes they enter. If community policing represents a threat to such self-understandings, it must be resisted, and strongly.

The police's sense of themselves as authoritative is important not just for how they define themselves vis-à-vis the citizenry, but also for how they define themselves to each other. Like workers everywhere, police officers seek to show their capabilities to their peers. But their efforts take on a particularly individualist cast. Officers hope to demonstrate their worth through their own idiosyncratic exercise of discretionary judgment. These organizational emphases on discretion and specialization are reinforced regularly, with considerable consequences for how officers act on issues of public concern. In such fashion, the project of community policing is further thwarted.

"I Need to Take Ownership": Specialization, Autonomy, and Organizational Disarray

One of the community police team members with whom I rode was still learning the ropes after just a few months on the job. This became apparent when I asked about his relations with other police units. He told a story about a memo he wrote to the detective unit in his first weeks. He used the memo to outline what he saw as the major problems in his area. He asked for information about any individuals the detectives suspected of creating these problems. There was no response. The officer said he interpreted this to mean that he needed to "take ownership" of these problems. He inferred that the detectives regarded the memo as evidence that he was shirking his responsibilities. He heard them communicate, albeit passively, a powerful message that he was on his own.

On another afternoon, I rode with a patrol officer whose beat included Blufftop, the large-scale public housing facility. The streets of the facility were frequently dotted with abandoned and improperly parked cars. As we cruised the meandering streets, he noticed an abandoned car he had previously ticketed for a tow. He was frustrated that the tow had not yet occurred. I asked him about his strategy for the broader problem of abandoned cars. He told me that he issued tickets when he could. He said nothing about coordinating any such strategy with other patrol officers who worked the beat or with the facility's community police officer. This was particularly striking because this officer formerly worked in the community policing unit and was a devout believer in the theory of "broken windows." In other words, this officer saw abandoned cars as important symbols of neighborhood decline that would invite more serious crime. He was also presumably well versed in the importance of coordinated action to address an intractable problem. Yet he, too, was flying solo.

Recall that community policing includes an emphasis on solving these more entrenched problems. Police-citizen partnerships should develop both a deep understanding of ongoing concerns and a thorough set of strategies for resolving them. As we have seen, the police are reluctant to surrender the prestige and moral standing they associate with authoritative and professional crime fighting. But they are equally reluctant to come together to address issues like abandoned cars in a coherent and comprehensive way. Implicit here is a valorization of individual autonomy and capability, a respect for any given officer's discretionary authority, a recognition that one must "take ownership" of one's responsibilities. A consequence is a frequent lack of coordinated action. As one officer told me, "Around here, one hand does not know what the other is doing."[12]

One obvious manifestation of this organizational disarray is the strained relation between the patrol operation and the community policing team. I saw little evidence that productive communication occurred across this bureaucratic division. Some patrol officers said they carried business cards of the community police team officer assigned to their beat and distributed them to residents who complained of an ongoing problem. Others said they often acted upon requests by community police officers to engage in extra patrol of a particular block or street corner suspected of hosting criminal activity. But never did any officer—from either patrol or the community police unit—describe any thoroughgoing, cooperative effort to address the underlying causes of an intractable problem. Indeed, a few patrol officers could not even name the community police officer who worked their beat.

One patrol sergeant with whom I rode had previously worked in a community police unit. He described the community police operation as a "garbage can" into which patrol officers dumped various nuisances they considered not worth their time. Patrol officers wished to focus on "serious crime," and saw community police work as trivial. One patrol officer with community police experience complained that a typical patrol officer did not understand what community police officers actually did. As a consequence, he acknowledged, there was very little sharing of information between the two units.

For their part, patrol officers operated largely as autonomous units. An exaggerated instance of this was the patrol officer who spent much of the time I rode with him outside the beat to which he was assigned. He focused on one particular house he believed was the site of ongoing drug activity. He knew many of the residents and talked regularly with the landlord. He obviously wanted to show me his in-depth understanding of the house and to document his efforts to reduce its capacity to host criminal activity. But he also made clear that his efforts were wholly his own.

Another patrol officer told me that he might refer a matter to the community police unit, such as long-term tensions between apartment residents and their landlord. However, if he gathered any intelligence about a hub of suspected criminal activity, he did not share it. This would amount, in his terms, to "passing the buck." He strongly preferred to investigate on his own. That was the best way, he said, to "learn things" and to take effective action.

Part of the issue here is the bureaucratic division of labor, and the manner by which it limits officers' focus. For instance, veteran officers often complain that younger officers are much less group-oriented than their predecessors. Gone, they suggest, is a greater sense of collective responsibility for monitoring and solving problems. They attribute this to the progressive parceling of responsibilities into specialized units. Indeed, community policing is cited as one instance of this. Some veterans believe that the creation of a separate community police unit has reduced the sense of responsibility patrol officers feel for their beats.

Beyond these bureaucratic barriers, however, lies a belief that credit and competence are individually earned, and that officers should spend their discretionary time as they wish. This was most obvious during my ride-alongs with three different officers with less than a year's experience. Two officers patrolled their beats with little discernible purpose;

they displayed minimal knowledge of their beats, they articulated no explanation for how they spent their time. They lacked direction, both figuratively and literally; they cruised without a goal, and they frequently got lost trying to reach the locations to which they were dispatched.

By contrast, the third rookie officer possessed a strong focus. He spent much time parked at a gas station in a poor neighborhood. As cars came in for fueling, he focused on those that were particularly run down. If he spotted such a car, he would input its license plate into his in-car data terminal. He searched, in particular, for information that the car's operator possessed a suspended driver's license. Such an offense in Seattle is subject to an immediate fine and an impounding of the car. The officer enjoyed this activity because he was often able to ticket and tow one or more vehicles during each shift. Regardless of his sense of accomplishment, the striking fact was that he acted on his own; no sergeant either endorsed or supervised his work. No apparent effort had been made to determine if this was, in fact, the most productive and effective use of his time.[13]

Indeed, officers had their own idiosyncratic preferences for how to handle time not devoted to responding to calls. One officer took a great interest in juveniles. He learned the names of many of the teenagers in his area and tried to determine if any of them were in trouble or headed for it. Occasionally, he got other officers involved in these efforts, once during a night when I accompanied another officer. The officer with the juvenile focus wanted to visit a home where he thought he might locate a runaway girl. The officer with whom I rode told me that his colleague frequently requested such assistance. He used this incident to describe the extent to which he and the other officers were given free rein. He listed other officers working that night and described their tendencies. These ranged from a pair of women who patrolled in high-crime areas to a senior officer who looked for places to nap. Another officer with whom I rode informed me that he enjoyed proactive police work. However, he could name only a few locations where he concentrated any such efforts. He visited none of them during the four hours I spent with him.

In sum, the police define authority and competence in largely individualist terms. Despite the tight sense of fraternity within the organization, fueled by collective concerns about outside meddlers and by the need to protect one another when danger arises, officers demonstrated an inability to develop collective strategies to address longstanding problems. Patrol officers rarely interacted with their community police team colleagues, and individual officers largely acted as lone wolves in whatever proactive strategies they did adopt. There was a striking lack of concerted, coordinated effort to confront the full complexity of those instances of crime and disorder that generated consistent complaints.

In short, perhaps the best place for the police to build community is among themselves.

The "Bad Apple" Narrative and the Marginalization of the Citizenry

Various emphases within the culture of the police, then, help explain why officers resist public input and oversight, and why they act with minimal cohesion in addressing ongoing problems. The narrative officers employ to explain criminal behavior is an additional factor that helps minimize police-community interaction. The police tend to explain crime as acts committed by a select number of "bad apples." This narrative reinforces the police's image as authoritative law enforcers and helps downplay any significant role for the community in meaningfully assisting with problem-solving efforts.[14]

Recall again the vignette that introduced chapter 3, of the lieutenant who chastised the justice system for not dealing aggressively with young car thieves. His solution to crime was obvious: find the bad actors and punish them through ostracism. Such thinking is hegemonic within the organization. When asked to explain why one neighborhood experienced a high number of calls for service, an officer asked me: "Do you want the actual reason, or the politically correct reason?" The "actual reason" was that the area was home to an unusually large number of people who received federal housing assistance through the Section 8 program. The officer believed that these impoverished people were more likely to offend, and thus that their clustering generated a localized crime problem. The way forward, for him, was simple—remove them from the midst of the otherwise peaceful community. Or, as another officer summarized it, "Crime disappears when the bad guys are in jail."

Indeed, catching "bad guys" is the principal motivation of many officers. This informs their geographies. One officer, for example, explained that he looked for places to patrol that were "target rich." He sought locales where he could find criminal behavior and acquire enough evidence to justify an arrest. One of his favorite places was a strip of cheap motels where small-scale drug dealing and prostitution were commonplace. Because these activities were often visible, the officer usually acquired enough probable cause to investigate further. This often yielded an arrest. This same desire for an easy citation motivated the officer described earlier who set up in the gas station looking for drivers with suspended licenses. Another officer described the street on which that gas station was located as an "easy" place to find evidence of criminal wrongdoing, and thus to secure an arrest.

This desire to isolate and remove bad apples from an area comports with the moralistic understanding officers develop to understand their work. Police work involves a cleansing of communities, a removal of the polluting stain of criminality. For this reason, officers often celebrate an arrest; for many, a day's work is incomplete without one. This helps explain why the community police officers embraced the short-lived "power community policing" model: it enabled them to experience the moral victory of an arrest, a far more satisfying prospect than an ongoing series of inconclusive meetings. And this narrative of morally necessary expulsion coheres with the "broken windows" ideology frequently employed to justify a range of police tactics. In practice, broken windows are people, and fixing broken windows means arresting them.[15] For instance, many officers focus on those who loiter around areas of suspected drug activity. Enforcing statutes on loitering is easier than enforcing statutes on drug delivery, so officers favor the loitering statutes.[16] These statutes also comport with the bad apple narrative and its emphasis on expulsion.

Two implications of this narrative deserve stress. One is the manner by which it leads officers to discredit other components of the criminal justice system, primarily the courts. If expulsion is necessary, then detention must be lengthy. Because the police can only arrest, they depend on courts to impose stiff sentences. They seethe when this fails to occur.[17] When one officer was asked what members of the community could do to help reduce crime, his message was simple and forceful: elect politicians who will require judges to send criminals up the river for a long time.

This officer's statement illustrates the second implication of the bad apple narrative: a diminished role for the citizenry in helping reduce crime and disorder. Notice that the officer suggests that community members act most critically as voters who only indirectly affect the police through the political officials they elect. The officer seems not even to entertain the possibility that a more direct connection between citizen and police is possible. The role he envisions for the community is rather passive.

Implicit in the bad apple narrative is an infected community. While one could potentially attribute a role to community dynamics in allowing or promoting the infection, officers downplay this possibility. Instead, an emphasis on the moral stain of evildoers leads officers to stress the need for expulsion. This valorizes police action over community action, because only the police—with their tactics, bravery, and coercive tools—can surgically extract the bad actors. The community is thus demoted to a mere information provider, as the "eyes and ears" of the police.[18] Even the patrol officer with prior work on the community policing team indicated that the most critical role for the citizenry is to provide complaints. Such complaints, he argued, make evident where a problem is emerging, thereby helping to galvanize a police response. The role for the citizenry is thus to help officers isolate wrongdoers, and then to step back when the cavalry arrives.

In this way, a notion of citizen agency—of citizens as co-equal partners in developing and enacting overarching strategies—is diminished. Once again, police superiority emerges as the dominant narrative, even if presented as the virtuous extraction of cancerous evil. Perhaps the public can point out the bad apples, but shaking them out of the tree is strictly a police matter. And even if the bad apple narrative possesses some verity, there is little consideration of any wider dynamics that explain their presence in a particular area, and little role for working with the citizenry to comprehend and address these dynamics. When the police employ this narrative of crime and its possible eradication they downplay the robust role for citizen involvement that community policing potentially represents.

Conclusion

Bittner correctly emphasized the centrality of coercive force to the structure and organization of the police. The capacity for such force is largely what makes the police a public institution of such symbolic significance. It also helps explain why the tensions between subservience, separation, and generativity are so vividly exemplified through an analysis of the police-community relation. Given their coercive authority, it is necessary that the police fall under public sway; the risk of an emergent police state ensures that the narrative of police subservience to citizen oversight will forever retain power. However, the capacity for coercive force also means that excessive subservience might leave the police susceptible to unwarranted uses by a particular social group. The liberal preoccupation with state neutrality, and its attendant emphasis on the regulation of the police through law, is quite understandable. Such legally produced neutrality should help ensure that the police do not suppress a disfavored minority group. But the capacity for coercive force also helps generate the moral framework in which the police operate, and through which they construct their epistemologies of community and their preferred relation to the citizenry.

My discussion here explicates further why the projects of separation and generativity stand in general tension with the ideal of subservience. This tension is exemplified in ongoing officer resistance to community policing. The ability to exercise coercive force is implicated in the cultural world officers construct, specifically in their emphases on masculinist adventurousness, safety, and authority. Each of these reinforces a notion of police separation from the citizenry, although not in keeping necessarily with liberalism's stress on state neutrality. Their sense of separation derives instead from their self-construction as competent professionals. Police understand themselves as superior to the untrained public and arrogate to themselves the right to act forcefully and authoritatively. In so doing, they obscure notions of a citizenry with a robust degree of political agency. The police also obstruct a more active role for the citizenry via their heavy reliance on a morally laden narrative of crime causation that emphasizes the ill effects of selected bad apples. This narrative implicitly trumpets the police as the agency of extraction. The citizenry is diminished as mere providers of information for police-generated tactics.

Even when the police are made aware of issues that community members wish to see addressed, they are unlikely to develop a coordinated strategy in response. A cultural emphasis on individual discretion and autonomy, and a bureaucratically constructed sense of specialized authority, work to keep officers in general isolation from one another. Collective efforts by the police to formulate a comprehensive approach to any problem are uncommon.

All of these police dynamics help shape how police relate to the citizenry. They also shape how the citizenry understands and evaluates the police. I use chapter 5 to examine how citizens assess these dynamics within the police force, and between the police and themselves.

"It Is So Difficult": The Complicated Pathways of Police-Community Relations

Lane was a longtime community activist who focused primarily on issues of crime and disorder. As well as any one, she understood the myriad challenges facing any neighborhood that sought to address and solve its problems. Because of her work, she was abreast of most neighborhood activities aimed at crime reduction, and she knew a great deal about the various bureaucratic agencies that could provide assistance. One of the big challenges, she stressed, was trying to mobilize those bureaucracies in a coordinated and effective fashion. Because crime problems were often shaped by other dynamics—by substandard housing, poor management of tenants by landlords, poor lighting and sanitation—she recognized the need for a comprehensive approach. This necessarily involved many agencies, and consequently entailed many headaches. As she described it:

Well, it's not strictly a policing matter, and that's where you get into so much trouble. It's not like a criminal violation where someone could be arrested and it's all over. It's just so much more complex than that. It usually involves so many different departments and so the jurisdictions of different people, trying to actually get at the heart of the problem is so difficult. And that's, I think, why it tends to have a real downward spiral effect on a neighborhood, because the people who really care about those things, they'll tend to give up after a while, because it is . . . it is so difficult to get any real long-term change.

Difficult indeed. In the preceding chapters I demonstrate that, by and large, neighborhoods that organize as "communities" cannot bear the weight that projects like community policing place upon them, nor do they wish to do so. I demonstrate further that, by and large, the police do not recognize neighborhood groups as co-equal partners in efforts to reduce crime and disorder. I also establish that ideals for constructing a police-community relation—as one instance of the state-society relation—come in three distinguishable yet sometimes conflicting versions. Given these realities, it is unsurprising that citizens in West Seattle found so frustrating their inability to move forward with the police and other state agencies on issues that concerned them. Like Lane, many of them understand why citizens tend to "give up after a while."

In this chapter, I explore how residents understand and evaluate the role of the police, and more broadly, the role of the state. There is no simple narrative that captures these assessments. There was considerable support for the police, yet much criticism; widespread understanding of the role and challenges of the police in society, yet much confusion about just what the police could and could not do. There was praise, honor, frustration, confusion, resistance—often within the same interview.

One instructive way of apprehending these myriad narratives is through examining the three main registers of state-society relations: subservience, separation, and generativity. In what follows, I explain how each of these informs citizen assessment of the police. At times citizens embrace each of these approaches, or at least express acceptance of them. Yet they exhibit frustration when these approaches are in conflict. Most notably, they describe how their desire for a more subservient state is frustrated in actual practice. Indeed, the principal source of citizens' discontent is their sense that the police are not properly responsive to public requests and suggestions. Many citizens suggest that the police are too aloof and too concerned with maintaining their authority. Further, residents complain that the moral frameworks and bureaucratic routines through which their input is channeled can be demeaning, limiting, and confusing. In other words, when officers distinguish themselves from the community, and when they construct the community, they often frustrate citizens' efforts to exert greater influence over what the police do.

At the same time, those interviewed expressed great appreciation for the work of the police, and an invariant acceptance of the police's role in society. No one, for instance, wished to abandon the police as the principal agent of crime control. Indeed, one of the most common citizen complaints was that the police were not present *enough* in their neighborhoods.[1] Yet these citizens struggled to see a way clear to a better reality, in part because of the conflict between modes of state-society relation. In short, the narratives of subservience, separation, and generativity alternately inform and frustrate the public in their quest for an ideal relation to the police.

I demonstrate the trenchant nature of this reality in what follows. In the first section, I review the broad acceptance citizens express for the role of the police, and the remarkable degree of empathy they feel for the challenges officers face. In the second section, I outline how each of the dominant understandings of the state-society relation resonates with the citizenry, how it is that residents simultaneously accept subservience, separation, and generativity. But this fundamental acceptance of the state, and of the possible ways of constructing the state-society relation, is complicated by the fact that these narratives are in complex and sometimes contentious relation to one another. This is the focus of my third section. The analysis there makes plain that the challenge of bringing society and state together in projects of neighborhood betterment is likely to continue to be "so difficult" to accomplish.

"It's a Tough Job": Citizen Acceptance of the Police

Given the actual and symbolic threat that crime represents, it is hardly surprising that many urban residents express strongly held views on the police. Indeed, they principally understand the police as their protector against such threats and evaluate them accordingly. The citizenry recognize the central role of the state in the suppression of crime and see the police as the principal agent in this effort. Never did a respondent suggest an alternate primary role for the police, never did anyone describe a vision of how to restructure what officers do. The centrality of the police in society, particularly in terms of an effective response to crime, was unquestioned. There exists a basic, if often implicit, acceptance of the state and its institutions of formal social control as indispensable tools in social betterment.[2]

One resident, for example, described her efforts to help minimize the negative impact of a young man whom she and others suspected of dealing drugs out of his house. She worked closely with a crime prevention official associated with the police department. As she recounted the various bits of advice this official provided her, she said, "He tells us to call the police. Who else is there to call? I mean, who else would you call?"

This resident is unusual in her explicitness, but her inability to imagine serious crime reduction without the police was commonplace. Citizens so recognize the police's significance that they regularly voice a desire to see more officers in their neighborhoods. As Marsha, a Midlands resident, put it: "I'd love to see a lot more patrolling . . . I'd love to see a patrol car down our block four times a day." Because she knew that some of the results of this research were going to be shared with the police department, she used the end of her interview to reinforce this point:

INTERVIEWER: That's all my questions. Is there anything else you think I should have asked or that I should know about?

MARSHA: No, I just think that you should know that we need more police. And whoever this report goes to, that's my biggest thing: we need more police.

Marsha was hardly alone in this request.

Residents thus accept the police's centrality and frequently express a desire for a greater police presence in their neighborhoods. Beyond this hegemonic recognition that the police were indispensable to efforts to reduce crime, many residents expressed sincere admiration. For example, many were quick to express appreciation for instances where officers responded effectively and sensitively. Among the teenagers interviewed, reservations about the police commonly emerged. However, one young man, Christopher, provided a poignant contrast. When asked to describe his evaluation of the police, he said this:

CHRISTOPHER: I like them, 'cause of what's happened in the past with my mom and her ex-boyfriend.

INTERVIEWER: Can you tell me about that a little bit?

CHRISTOPHER: OK. My mom's ex-boyfriend, she had let him live with us, because he had lived with his mom. And at first it was OK, but then, my mom and him went out one night—they went to drink, you know, and he hit her, and then they ended up going home. . . . It was like during the day, when I went to the mall with one of my friends. And he hit her, when I had came home, he started slapping her and stuff. So I called the police, you know, and ever since then I trust the police, I like them more.

INTERVIEWER: Cause they handled it well?

CHRISTOPHER: Yeah.

INTERVIEWER: In what way did they handle it well?

CHRISTOPHER: Like, they took my statement and my mom's statement, and every time the phone rang, and I answered it, they tracked him down and took him to jail. So I trust them more.

Another of our interviewees was troubled by a friend whom she had invited to stay with her temporarily. When her friend began displaying symptoms of emotional distress, the living situation became tense. Ultimately, the woman called the police to help her remove her friend from her home. Throughout the process, the woman said, the police were sensitive to the dynamics of the situation and helped resolve it peaceably. This led her to praise the officers' work:

And they were very helpful and comforting, and were easy on me—gave me a feeling of being supported. In other words, they didn't try to blame me for the situation, even though they could have questioned my side. . . . So, I felt supported, and everything resolved very quickly and easily, in terms of getting him to leave and stuff like that.

Although this respondent's interaction with the police involved an unusually sensitive matter, her appreciation for quality police work was echoed by many other respondents.

Residents also frequently expressed empathy for police officers. Many said they understood the difficulty of the job, and recognized various impediments to what officers could hope to accomplish. One of the more eloquent expressions of these sentiments came from Sherry, a resident quite active in crime reduction efforts:

You know, we really put a lot on our police, we really do. We expect them to baby-sit, we expect them to be psychologists, we expect them to use force when necessary. I mean, I have seen the police take a lot of abuse and you know not do anything. I have seen that actually happen. . . . So it's a tough job, it's really hard being a police officer in the world in which we live today. There are a lot of expectations.

Sherry recognized how officers might struggle to satisfy a public with perhaps excessive expectations. Given these struggles, she indicated strong appreciation for officers who did well. But, she said, "I also know officers who should go do something else." Although expressed here in the negative, even this statement recognizes that the police carry a welter of expectations, a burden so heavy that some officers cannot bear it. Or, as one resident put it, "I feel sorry for them. I would never be an officer, never."

These comments demonstrate that many citizens so accept the role of the police that they are able to imagine the challenges officers face. Such comments also show that many recognize that citizen expectations of the police are sometimes excessive. Implicit here is a realization that perhaps the police can only do so much to improve neighborhoods. In a discussion of the scope of the police role in bettering communities, Marshall, the pastor of a church in Centralia, said this:

I think the police have to play a role, but maybe the reason I'm saying that, that there are other ways of doing it besides them, that we need to look at. They are busy enough. I mean, hey, they keep me safe at night, I appreciate them for doing that. If they want to focus on that and cleaning up the streets, I say you concentrate on that and let us, and find other agencies or whatever, or have the police come alongside churches that can help change the community. But you put the police in charge of changing the community as well as policing the community, they are going to fail in both areas.

And even in terms of reducing crime, many residents were quick to recognize limitations on what the police can do. In colorful language, Rob, the activist in Centralia, made this point: "And I say, we have lost the battle when we pick up the phone and say, '911, we have a crime happening.' And a cop comes in his 911 car. Right there, you're on the way to the septic tank." For Rob, reactive police work was of only minimal significance. The more important work needed to occur beforehand, to create neighborhood conditions in which crime did not flourish. For this, as both of these respondents indicated, the police were of no significance.

In sum, residents typically: accept a central role for the police in society; appreciate quality service even while recognizing the myriad challenges officers face; and recognize limits on what the police can hope to accomplish. My larger aim here is to demonstrate that citizens see the state's role in crime reduction as a legitimate one; there is hegemonic acceptance of the state's need to equip and make available a police force. There is even widespread acceptance of the three means by which state-society relations can be established, although this is entirely implicit.

"Oh, Look at the Police!" The Resonance of Subservience, Separation, and Generativity

In chapter 3 I showed that the three main frames for state-society relations—subservience, separation, and generativity—all shape how the police approach the citizenry. In many instances, the police demonstrate an acute desire to respond effectively to citizens' requests. However, they often reinforce a self-construction as an aloof and authoritative force. And, in all instances, they act upon an implicit sense of what "community" is, and act through a set of bureaucratic routines. So, even if these three means of constructing the state-society relation are in frequent tension with one another, each is still potent in structuring police relations with the communities they serve.

Similarly, all three state-society paradigms influence how the citizenry understand the police. The interviews revealed acceptance of the legitimacy of each.

Subservience

Not surprisingly, citizens commonly endorse the idea that the police should be subservient to the public. Residents most frequently made this plain when discussing their requests for police service. When residents call the police, they expect to be understood and to receive an indication of how the police will respond. The interviewees regularly stated a need for the police to be more available and responsive. This emerged typically as our respondents described the *absence* of a desired police response. Residents were dismayed that requests for service generated no visits from officers, that phone calls to community police officers were not returned. They were often frustrated that 911 operators did not understand what they wished to report, that patrol officers were slow to show up or never arrived at all. They expressed frequent dissatisfaction with a police force that seemed incapable of making explicit what the public could expect from a call for service.

So, the public expects a subservient police force primarily through faster and more thorough responsiveness. But for many, a desire for something more was evident. They hoped for a deeper sense of connection to the police than they regularly experienced. Part of this seemed to be a nostalgia for a perhaps imagined past when the officer on the block was a familiar and central figure in community life. But, more typically, the justification for a closer connection was to improve the flow of communication. In this way, many respondents suggested, the police could be more responsive to the community's input. Terry, a Centralia resident, was asked what would make his neighborhood more of a community.

TERRY: Well, I think if we had six or five police officers like this walking in the street, like getting to know the neighbors, their names . . . kind of like the people that are most troublemakers like neighbors . . . you know, the neighbors [that are] the most troublemakers, stuff like, like. . . .

INTERVIEWER: That would help make it more of a community then?

TERRY: Yeah, so it would be like safer and people wouldn't be so much scared of the police any more. 'Cause some of the people they would be like, "Oh, look at police!"

Another resident, Mark, expounded on the same theme in his desire for not just more officers, but more approachable officers: "I've heard it said, 'that I wish there would be a patrol car moving up and down the street.' But I think that that patrolman could get out of that car and actually walk the neighborhood. I think they would be able to be a little bit closer to the community." This closeness, he later elaborated, would help the police understand more completely the sources of residents' concern. Officers would then be better positioned to respond well to citizen requests. Some focused specifically on teenagers. They theorized that teenagers often know a lot about existing and impending criminal activity. If the police could establish better ties to teenagers, they would acquire more data about where crime might emerge, and could plan accordingly. Richard, from Westside, explained this point well:

And you know what would be a real plus, is if the police department would be little bit more friendly towards some of the kids around here. Not all of them. But I think a lot of them are put in the same basket. The bad and the good are all mixed together. And I think that they assume that all the kids are bad. Or are going to be, or doing something suspicious or something awful. And a lot of them aren't. But if they actually could take a couple extra seconds and stop and talk to some of these kids, at least some of the older ones, it would be a real plus. And it would be great for P.R. And it would pick up a ton of information.

Over all, it was striking that residents so regularly evinced a desire to know the police on more informal terms, to establish a chattier relation to a regular patrol cop. Because many of them believed that a more informal relationship with the police would increase their access, they endorsed a closer tie to promote greater responsiveness and effectiveness.

By contrast, however, none suggested a need for the public to oversee the work of the police in any meaningful fashion. There were no endorsements for a police review board or any other more formalized means by which the public could evaluate officers. Rare was the interviewee who expressed no interest in police reform of some sort; more commonly, there were numerous suggestions for how things could change (about

117

which more below). But even those who were most frustrated never advocated formal mechanisms for citizen oversight of the police. Granted, few respondents complained about excessive use of police force. More such residents in the sample might have revealed greater interest in an institution such as a citizen review board. Even so, during the research period, SPD officers shot and killed three citizens in circumstances sufficiently cloudy that activist groups complained bitterly about police brutality and used local media to press for citizen oversight. In such a historical moment, it is striking that none of those interviewed seemed persuaded that a move toward formalized citizen oversight was necessary.

In short, the oft-stated desire for more, and more responsive, police officers illustrated the profound resonance of the narrative of subservience. Citizens wish to relate well enough with the police to influence what officers do, and they expect a response to a call for assistance. And many of them desire a connection of some familiarity and casual conversation, to enable a flow of information that can equip the police to solve the problems about which residents complain.

But even if many wished this informal connection, many simultaneously recognized that the police must, at times, maintain their distance. In other words, they recognized the legitimacy of police efforts to position themselves as separate from the public. This push toward separation has two sources, both of which were recognized by the citizenry. They saw that the police must obey the dictates of the law even in the face of public demand. They also sometimes applauded officers who maintained an air of professional authority.

Separation

As much as residents might wish a police force that served at their beck and call, they recognized that officers at times need to separate themselves from the citizenry and its demands. For some, this police effort at distinction was understood precisely in the terms of classic liberal thought—as a need for officers of the state to remain loyal to a more abstract legal code. Such loyalty means that state agents avoid allowing the parochial agendas of one social group to override the rights of more marginalized groups. In this way, the ideal of equality is ostensibly served; in following the law in like fashion across the social landscape, the police should treat all the same.

Residents demonstrated an acceptance of this legally mandated separation between the police and the citizenry in two ways. Some expressed a deep empathy for the bind created for officers by the tension between subservience and separation. Meg, an African American in Centralia, captured this well when discussing how many of her neighbors hold the police in contempt:

I don't think that any of us out here should be involved and dealing with police, if you have no knowledge of what does a police do, what is his job, what is expected of the police. If the police doesn't follow through with their job, there's a process for that. You know what I mean? But to sit up and kowtow and all that kind of stuff, and always on the police's back, I'm just not like that. I give everybody the benefit of the doubt. If you're a police officer and you screw up, I'm gonna tell you, you're a screw up. I'm not going to go out here beatin' a drum and generalize to "the police." 'Cause that's not fair to the ones that's out there that's totally dedicated, totally doing their job and trying to be Officer Friendly. Some of these people, they experience terrible lives because they're police officers. It's not easy on them, that's not an easy job. You're a police officer, and you have to stop and think of the boundaries of the law and dealing with this chaos situation you dealing with at the same time. They have to think twice as fast as the rest of us!

Meg's degree of empathy for the police is unusual for her neighborhood. She recognizes that adherence to legal procedures can often challenge officers dealing with a demanding public in often chaotic situations. Note that the law is relevant because of the "boundaries" it places around permissible police behavior. Such boundaries codify the separation between the police and the public.

The recognition that adherence to the law limited police action was most readily expressed by those active in crime reduction efforts. Such activity necessarily placed citizens in closer contact with the police. They were thus more likely to be told how enforcement strategies could not ignore legal proscriptions. Recall, for instance, the wooded area of a park reputed to host anonymous sexual encounters. This phenomenon was repeatedly raised at community meetings. Each time, the community police officer reviewed how making arrests required a level of evidence his surveillance tactics were unlikely to yield; the law limited his ability to meet citizen demand. Those who conversed more regularly with the police learned of other instances where this was true, and thus generally expressed greater acceptance of how separation often trumped subservience. For instance, Nora, an Eastside resident, had attended the citizen's academy established by the police department. This six-week program schooled citizens on key issues of police practice and procedure. Nora explained why she found the program helpful:

I think that it helped me to understand that it is an incredibly challenging thing to do, and really difficult, and that it requires a lot of patience from the individuals—the police officers themselves—and that it's just a really challenging thing to do because you are kind of between a rock and a hard place with the demands of the community or an angry citizen who is in your face, and knowing that you have to follow the letter of the law, whether or not you think it's best or not.

Residents also expressed acceptance of legally created separation by endorsing the ideal of equal treatment that such separation is meant to accomplish. Most commonly, this emerged as residents complained about their perception that their neighborhoods were slighted. Many believed that the political clout of more affluent neighborhoods translated into higher levels of police service. Their indignation at this perceived reality revealed their belief that equal treatment under the law demanded greater observance, that some communities' ability to make the police subservient limited the fairness that separation ostensibly promoted.

Many residents also recognized that the police's internal drive toward separation—one based on adherence to their professional norms—was legitimate. Many openly accepted the argument that the police were justified, at least occasionally, in exuding an air of authority. Recall that officers endorse this practice to assert control over potentially chaotic and dangerous situations. Some of those interviewed accepted the police's justification for this practice. Said Geraldine, a Blufftop resident who was interviewed jointly with a neighbor:

I think that they have to have that element, like playing poker—they have the ace, they have the better hand. They have to make their showing that they're not just a pushover, because they are there to keep the law and order of the land, and they would say, "That's what we're here to do."

At this point, her co-interviewee, Ashley, chimed in: "That's what we pay them to do." For these two women, as for others, police displays of authoritative separation were functionally necessary to enforce the law and maintain order.

Another respondent, Roland, recognized police displays of stern authority as something of a self-protective mechanism. As he put it:

But like I said, they deal with society's b.s. all the time. They see wife beaters, and child killers and rapists, and they deal with the b.s., so you've got to put on an authoritarian attitude, 'cause if you're just the same person you're at every day, you're gonna get changed.

In these ways, many of those interviewed expressed an implicit acceptance of the narrative of separation between state and society. They recognized that adherence to the law meant that the police could not always comply with public demands, and that the maintenance of authority helped officers succeed and survive. As is elaborated below, however, the extent to which police officers embrace this narrative of separation frequently frustrates citizens. So, even if residents recognize the legitimacy of separation, they question how it can inhibit police acquiescence to citizen requests. A similar ambiguity is attached to generativity.

Generativity

Although they are in frequent tension, the narratives of subservience and separation both presume a strong division between the police and the citizenry, between state and society. In practice, however, the distinction is often blurred, because the state fundamentally constructs the community with which it interacts. I demonstrate in chapter 3 that this occurs in two fundamental ways in the case of the police. One involves the bureaucratic routines and other cultural practices by which the police apprehend and act upon public input. Via these routines and practices, the police determine what "community" actually means in practice. Generativity also operates through the moral understandings officers develop for their work. When they couch fighting crime and protecting the vulnerable as moral crusades against an evil that demands expulsion, police officers obliterate the line between state and society. Police and community are joined as twin forces of good, united in an effort to remove the stain of criminal pollution.

Many residents implicitly recognized the salience of each of these generative operations. Although the bureaucratic routines through which citizen input was channeled often prompted complaints, there was a simultaneous recognition that such routines were unavoidable, and even desirable. This was more commonly expressed by those with greater awareness of police procedure, such as those who had attended the citizen's academy. One such resident was Lane, the activist quoted at the beginning of this chapter. She found herself regularly schooling her fellow citizens on how the police operate. At more than one meeting, she explained how complaints need to be made, how the police categorize and prioritize the input they receive. She hoped that greater awareness of how the police bureaucracy apprehends complaints would enable others to mobilize the most effective response. But even those with less understanding recognized the need for procedures to differentiate between, for example, emergency and non-emergency requests for service, or for a specialized procedure for complaints about drug activity. They implicitly accepted the reality that the police structure citizen input via prescribed routines.

Many of them also mimicked the police by moralistically framing officers' work as a mission to expel criminal evil. Such a framing exalts the work of the police and places it above credible citizen critique. As Marshall, the pastor, put it:

I have a positive view for them, because I lay in bed at night while they are risking their lives so I can lay in bed at night and sleep. And if you want to talk bad about that, then you are the biggest idiot I have ever met. They are not perfect. . . . But hey, a guy that's going to risk his life and enjoys it, so that I can have a safe community to live in, kudos to them.

The moral standing of the police is made most evident, for Marshall, by the fact that officers stand to pay the ultimate sacrifice to protect the citizenry—a view officers themselves endorse. From this perspective, "talking bad" about the police is idiotic. Such morally upstanding public servants deserve the benefit of the doubt.

Although residents were not universally quick to view the police in such sanctified terms, many did frame crime as an evil in need of expulsion. And everyone saw the police as vital to that project. Indeed, the expulsion narrative's potency revealed itself even in matters that did not directly involve the police. In the most distressed neighborhood in the study area, Centralia, community activists, along with the police, developed a hegemonic explanation for the source of the crime problem: inattentive absentee landlords. Such landlords, the common narrative went, did not monitor their residents, most notably by rarely evicting those who caused trouble. The landlords were thereby ostensibly guilty of allowing criminal behavior to fester. This narrative is notable because activists broadened their anti-crime scope beyond the police. In fact, they largely absolved the police of ultimate responsibility for the crime problem. Yet the residents shared with the police the basic framing of crime and its solution. Crime is a problem perpetrated by select, morally problematic individuals who need to be expelled by a legally empowered agent of territorial force.

As with the prospect of separation, though, there were limits to how much the citizenry accepted the police's generation of community. Let us turn now to how the citizenry expressed resistance to separation and generativity.

"I Don't Know Why They Can't Smile": Separation as an Obstacle to a Responsive Police

In the abstract, the narratives of subservience and separation can be complementary. The term "liberal democracy" is not meant to be an oxymoron. The individual and collective rights that liberal regimes protect through adherence to legal rules should create a political space for the development of an informed and engaged citizenry. Such politically well-developed citizens can subsequently enact democracy in a responsible and effective manner. Without these rights, citizen activism cannot develop with sufficient distinction *from* the state, and thus cannot have meaningful purchase *upon* the state. The liberal project and the democratic project are necessarily joined.

In practice, however, they can conflict. As I noted above, the residents recognized the salience of both narratives; they wanted a responsive police force, yet they recognized the police's need to retain a distinct authority. But they regularly revealed considerable frustration with how the push to separation often blunted responsiveness. Some of this frustration was tied to the mandate to require that public agents be governed by a fair and objective legal order; citizens sometimes complained when the legal code restricted an ardent police response to a problem. But much more frequently, citizen anger was directed at the other source of separation–police efforts to establish themselves as an aloof and muscular force of authority.

Frustration with legal restrictions emerged most commonly in discussions of sites of longstanding criminal activity. For example, more than one neighborhood contained a home that both residents and police suspected of hosting sales of illegal drugs. Over time, as problems of late-night noise and frequent, erratic car traffic continued, residents grew frustrated with the lack of resolution. One couple, Rebecca and Frank, who lived next door to a home with multiple dilapidated cars on its lawn and much suspected drug activity, went to great lengths to document the problem. They recorded license plates of those cars that came and went, they used a video camera to document the late-night activities. In so doing, they incurred threats from their neighbor and his friends. After going to such great lengths and making themselves vulnerable to retaliation, the couple understandably wished for a forceful and effective police response. But the police could not act just as the couple wished, most significantly because of laws governing private property. To acquire a warrant to search the premises, the police needed strong evidence of drug dealing. In pursuit of that evidence, undercover detectives sought to purchase drugs in the house. They did not succeed, the couple was told, because the owner of the house would not sell to those he did not know well. The problem festered, as did the couple's frustration.

Such frustration was common in areas where similar situations existed. One activist summarized this sentiment in the neighborhood group in which she was active:

I know that we have some other folks that attend the [neighborhood group meetings] on a regular basis that were incredibly frustrated with how long it took the police to move on a drug house that everybody knew was a drug house. But because there's funny laws with ownership of the house and who is living in it, and the landlord thing, and it just took a long time to make anything happen with that, and . . . so I think that people are incredibly frustrated with how long it takes to see results.

The same dynamic emerged around the anonymous sex in the park. Nearby residents expressed displeasure with the legal obstacles that limited an aggressive police response.

However, this displeasure with the limits of the law on police enforcement practices emerged less frequently than commentary on the cultural practices of the police. Thus, it is this second source of the police's drive toward separation that attracts the most citizen ire. Even if residents recognized the functional need for officers to assume a posture of authority, they simultaneously resented the extent of such posturing. Frequent complaints emerged of officers who positioned themselves as too aloof and dominant a force and thereby repelled citizen interaction. Said one Midlands resident, Andrew:

Cops don't understand that they are kind of threatening, just really kind of menacing looking. . . . And some cops, they don't smile, they don't really interact. You almost feel like there's a wall between you and the uniformed police. . . . I think they would do a lot better if they just lighten up a little bit. I think they'd get more cooperation, if they were just a little less standoffish.

The metaphor of a wall was used with some frequency. Sally, also of Midlands, said:

So, I know that it has to be a really stressful job, and so some officers tend to wall themselves off, by just deciding that they're going to do their job, they're going to be alert—there's kind of this hyper-vigilance that they develop, where they're always trying to pay attention to movement and suspicious circumstances. And if you're only focused on that, and not on the people and not on community, I think it's easy to kind of lose focus, even if you were initially motivated to get into police work because you wanted to make a difference.

While Sally understands why officers might fall into a state of "hyper-vigilance," she simultaneously understands how this can hamper police-community relations. Or, as Betsy, a Midlands resident, put it, "Sometimes you see too much *NYPD Blue* and not enough of what policing really is, which is peacekeeping." For Betsy, a "peacekeeping" approach emphasizes not the force of police authority but more cordial relations built around collective problem-solving. She perceived too little of this cordiality, and she believed its absence was detrimental to police-citizen cooperation.

For many, then, police officers seemed to emphasize the dangerous aspects of their jobs, and thus asserted an air of authority to command every situation, even when no danger existed. For others, police authority was interpreted in terms of professionalism. The woman described above, Rebecca, whose neighbor's alleged drug sales the police could not abate, was annoyed by her interactions with the police. She detected a patronizing tone. Her language was strong. At one point in her interview, she described her general frustration: "The police don't want the citizen to be the eyes and ears for the police, in my experience, that it's seen as infringing on the territory of the police. They're the experts." She expanded on this point while describing her actual encounters with a community police officer:

[The community police officer] never said, "There's something bad over here, I'm going to try to help you." Never had that kind of relationship. "There, there, nice little girly. I will take care of this problem, you leave it to us, we're the professionals"—that's what I got. Then I'd get stonewall.

Even though Rebecca and her husband went to great lengths to assist the police by gathering significant evidence of criminal activity, they never experienced a sense of shared governance. Rather, the police's desire for professional stature seemed to prevent the cooperative relation the couple desired.

Many residents complained that this lack of cooperation was further thwarted by their neighborhood's lack of a regular patrol officer.

Instead, they described such an extensive parade of officers that residents had no chance to establish a connection with any one of them. Said one:

I wish they could have stayed around more. And I wish they could get to know people. Some of the police officers are pretty nice and friendly and they get to know people, but then you never see them again—they're gone. And people just get confident, to tell them or talk to them—you know, whether it be just general conversation or letting them know information, things that are going on. But by the time people get the confidence up to talk to them—approach them—they're gone. I mean, they are literally gone. They're moved somewhere else, or they . . . probably don't want to work in this area, so they ask for a transfer somewhere else. I'm not sure how it works, but they just don't stay.

Another resident, Jack, who had been active for several years in crime prevention efforts, complained that even he did not know more than two officers who worked in his neighborhood. This belied any perception that officers were genuinely interested in serving their neighborhoods and prevented the development of sufficiently close ties to enable a productive flow of information. One resident said she wanted to know that the police were "nice and friendly and someone you can go to for help." She indicated that she would be much more willing to engage such officers. Because she saw a "constant rotation" of different officers, she never developed the close relation she desired.

This sense of officers constantly rotating frustrated hopes for improvement. Said Meg:

We don't get to keep nobody. They change them police officers like you change underwear. As soon as get to know them, get to working with them, they move them on to another station somewhere. That doesn't help then either because then you have to keep starting from scratch.

For many citizens, then, the police were far too separate from the public, too difficult to engage because of their displays of authority and their transience. This air of distinction was made worse for some by the extent to which the police affected a particularly masculinist pose. Tracey explained:

Oh, I don't know why they have to come on like they are macho, macho power people, I don't know why they can't smile and, you know, give me a ticket, talk gently, they always come across, you know, maybe it's my impression of the police before they come to me, but they always seem to come macho, they never smile, and they make me feel like I'm just a terrible person.

Said Roland: "They're really abrasive. If it was some guy off the street acting that way without a badge, he might catch a fist in the mouth, but since he's got a badge on, it's sort of a crutch he can lean up against and act any way he wants to."

So, the expression of masculinist authority keeps many members of the public at significant distance from officers. This impedes the type of relationship that citizens say would allow them to influence what the police do, to make the police somehow subservient to public input.

Not surprisingly, excesses of masculinist authority were emphasized most by African Americans.[3] Indeed, the only detailed stories that emerged about the direct experience of rough handling by the police came from two African Americans. One, a teenaged male, detailed a scenario in which officers grabbed and searched him. The officers later justified their actions as motivated by a report that someone in his group had brandished a gun. For his part, the young man was not convinced that the police would have treated an older person in quite the same manner. Similarly, Lacey, an African-American resident of Blufftop, described an incident in which she said her son had been treated roughly. When asked if she had any unpleasant experiences with the police, she said:

LACEY: No, but my son has had encounters with them, and as far as I'm concerned, they get on my nerves. Because they have stopped him and accused him of being somebody else, and throwing him up on the car and handcuffing him and checking him, and you know what I'm saying. And they got him for jaywalking in front of the house.
INTERVIEWER: Why do you think that's happened to your son?
LACEY: Well, because he's black. And because he's 6 feet 5 inches tall. And because he was wearing a hooded sweatshirt, and it was the color of, I guess, gang members.

These complaints from some African-American respondents point to a particularly problematic form of police authority. But they were hardly unique in their general complaint about the narrative of police separation. Residents regularly suggested that assertions of police authority made officers unapproachable and even unreachable. As a consequence, they were unable to make one agency of the state, the police, sufficiently responsive. Subservience was thus impeded by the police's stance as a distinct and superior social group. Much the same dynamic reveals itself when we look closely at citizen reactions to how the police generate the communities with which they interact.

"There's Only So Much Self-Sacrificing You Can Do": Routines, Morality, and the Police-Community Connection

For the police, "community" is not an objective entity independent of their own social practices, but something they create via their internal procedures and cultural understandings. There are two principal means by which community is so generated: through the various processes by which citizen input is collected and sorted; and through the moral understandings officers develop to understand their work. In each case, the police construct "community" in quite consequential ways.

As I suggested above, many residents fully recognized the need for their requests for service to be channeled through various bureaucratic routines. Further, many of them share the moral construction of crime and of police work that officers favor. However, this basic acceptance of the police's generation of community was tempered by strong concerns. For many residents, the treatment of their requests and suggestions was a source of great frustration. Similarly, many expressed strong reservations about how police moralizing led to officers treating their concerns with insufficient seriousness.

Police protocols for assimilating citizen input emerged frequently in conversations with citizens. Many expressed deep frustration with how their conversations with the police were structured by a particular format. For example, those who called the police emergency number wished to give their story in their own terms. Instead, they were expected to answer a series of questions posed by the operator. As Lane put it, when people place a call to 911, their narratives are just "stream of consciousness." They find it jarring when the operator interrupts to ask a sequence of questions.[4] The caller's desire for empathy and patience is thereby disrupted. As someone who was well-versed in police procedure, Lane captured the dilemma well:

The problem is, though, in order to have optimum safety for the officer, is that they have a very specific protocol in the order of the questions they ask, if they are going to be dispatching an officer to the situation. And they can't wait for somebody to just disclose all these things. They have to ask all these questions.

Still, she argued that 911 operators needed to be more sensitive to the dynamics on the caller's end, and to respond with greater empathy. Such sensitivity would surely be welcomed by Jack and Lolly, a married couple in Eastside, who were quite active in neighborhood efforts to reduce crime. They described their frustrations with police response to their calls:

LOLLY: We have all taken notes and got notes and license plates numbers and. . . .

JACK: Not all of us, but those of us who are active in it, we'll try to collect information and report it, but. . . .

LOLLY: It's "What are they doing? How do you know? Are they doing anything?" And you feel like, why am I making this call? I've taken the license plate number . . . endangered my life, sometimes. . . .

JACK: . . . expressed a concern. But the 911, or the non-emergency, operators question the. . . .

LOLLY: . . . validity . . .

JACK: . . . of the report. They don't accept it. They question it. And we don't understand that.

When their input is channeled in a particular format, and questioned as to its accuracy, Jack and Lolly (and other residents) feel betrayed. Others were incredulous because they believed the questions they were asked evidenced a lack of awareness of the situation facing the caller. Two respondents were particularly appalled. Said Kristi:

And yes, we do call the police at 2 or 3 o'clock in the morning. We got people going up and down the street going 100 miles an hour. "Well, what's the license plate?" Who the hell knows?! I'm lucky to know that the car is blue. I am not going to go out there in my pajamas and stand and get killed to look for a license plate. So, if you can't give the police a license plate number, they're not coming out.

And Marsha:

I've called about gunshots at night, and stuff like that. They have said, "What does he look like, what kind of guns he's got, this and that." I'm not going to go outside asking that guy, "Can I ask you what kind of gun you got, what your intentions are in using it?" We can't do that, we are putting our lives at risk doing something like that.

Even when they recognized the need for the police to channel their input via certain proscribed routines, then, citizens were often disappointed when they encountered those routines. Operators who followed a script interfered with citizens' desire to tell their story on their terms, and thereby seemed to indicate indifference to the reality that callers faced.

This general sense that citizen input could not be heard went beyond encounters with 911 operators. It was also directed at non-emergency contacts with the police department, and with other municipal agencies. Many were regularly discouraged by the amount of effort required to make themselves heard. One woman, Patty, had administered care to a young shooting victim in Centralia, trying to stanch his blood flow while an ambulance was summoned. Though her efforts to save the young man's life were unsuccessful, Patty devoted herself to trying to prevent future such deaths by becoming involved in neighborhood anti-crime activities. She quickly became frustrated by the obstacles and difficulties involved in making her neighborhood's concerns heard and getting them addressed. She described some of these efforts and how they developed, and then complained: "But darned if the requirements aren't all self-sacrificing! There's only so much self-sacrificing you can do when you've got someone shot right in front of your house."

For many, this self-sacrificing took the form of trying in vain to locate the proper city bureaucracy from which to solicit assistance. One resident, for example, described the "voicemail maze" she confronted when she sought help with a run-down property on her block. The complexity of the task, she said, led her to simply give up after a while. For others, frustration emerged from the lengthy process through which citizen input is often vetted. Meeting follows meeting, and progress occurs slowly, if at all. Said Marshall:

And I've talked to the people in the community that used to go to the meetings but they don't go anymore for the same reasons. You have three people that talk all the time and they just argue with each other and it's about a light pole. Let me tell you, my life is too busy to talk about a light pole.

Or, as Rob said: "The process is a big production. I'm not really much of one for process. The process is way too much."

These and other residents feel deterred when they request a clear hearing of their complaints and ask for a sensible state response. The various bureaucratic routines through which those complaints are processed diminish citizens' confidence that they can render the state subservient.

Citizens are at times similarly ambivalent about the police's moralistic constructions, particularly of places. Many residents accept the police's moral construction of crime as an evil pollutant that requires expulsion, and many accept the attendant idea that the police are the virtuous agents of extraction. Tied to this moral construction of crime is often a moral construction of particular communities—as places where the pollution of crime is allowed to fester. Residents in these areas often experience the police as dismissive of the neighborhoods in which they live, and as neglectful of their problems. Jean lived in

Eastside, but she was active in anti-crime efforts in Centralia. She told a story about what transpired on an evening shortly before her interview. She and some other neighborhood activists were spending the evening barbecuing and keeping an eye out on the scene, in hopes of helping deter crime. She described what happened when the police came through on patrol:

Well, then the police came by, and it was three, no, two officers that we did not know—we see them on our block walks at night and we're starting to learn the familiar faces of the guys, but these were two that we didn't know. And they said, "What are you guys doing here?" And I said, "Well, we're just sitting here kind of watching life go by and kind of keeping an eye on the neighborhood. There's a lot of activity on this particular block tonight and we kind of just want them to know we're here, that the neighbors know that we're here, and we're kind of just observing things." And they says, "Gosh, we wouldn't sit out in this neighborhood at night by ourselves!" And I say, "This is our neighborhood! We can't just, you know, let it go by. We have to know what's going on." And then they kind of chuckle and they says, "Well, you'd better be careful that the drunk drivers don't come and roll over you here in the traffic circle." And we says, "You know, it's our neighborhood."

Rob complained about what he termed the police's "preconceived ideas about what a community is about," as part of an explanation of his dissatisfaction with levels of service from the police:

Well, they don't come for long hours, and when they do come, they're coming into a neighborhood where you're stupid to be living here. I've had them say stuff like that to me. "I wouldn't live in the city! I wouldn't do this." You don't feel protected by the police.

As members of an apparently morally condemned community, these residents suggest that they suffer from a lack of adequate service. Here, the moralism of the police seems to interfere with the residents' desires to make the police subservient to their expressed needs for a higher level of protection. Not surprisingly, many of these citizens suspect that neighborhoods of higher social and moral standing are not quite so starved for attention. Betsy was a landlord in a disadvantaged neighborhood down the hill from Blufftop. She had worked to rid her neighborhood of various instances of criminal behavior, but she felt less than fully supported. She suspected that if she lived elsewhere, her complaints might resonate more loudly:

BETSY: Because chances are, if I live in [affluent neighborhood] and I call in and say I've got an abandoned car here, somebody's going to come along and pick it up. We had eight abandoned cars on this cul-de-sac about six years ago. And it took a concentrated effort of a lot of residents working in concert with the crime prevention specialist and basically people just nagging and nagging and nagging until it got done.

INTERVIEWER: Why do you think it is that they leave them here longer?

BETSY: Because, they don't c . . . because we don't have the same amount of clout, economic clout that other parts of the city do have. Where we, when we do things effectively, it's because everybody gets together and just nags like little dogs, on getting it done.

It is difficult to verify the accuracy of Betsy's comparative assessment of police services.[5] But she made clear that she felt that the police were dismissive of her concerns because she lived in what they perceived to be a morally tainted neighborhood. Others felt similarly. Said Hal, a resident of Centralia: "When the police do respond, they come to the community with an attitude. Like, 'what are you doing living here? If you choose to live here, you're a loser.'"

Thus, the operations by which the police construct community—through their bureaucratic routines and their moral definitions of places and people—frustrate the desires of many residents to convince officers to provide the level of service they desire. As with police practices of separation, these generative operations run counter to the goal of subservience, in ways that lessen the legitimacy of the police.

Conclusion

Residents evinced both a strong acceptance of the police and an interest in working as closely as possible with officers. They saw the police as indispensable in helping make their neighborhoods safer and more secure. They yearned for a steady and reliable officer on the beat, one they could approach easily and engage in a productive discussion about their neighborhood. In this way, they implicitly endorsed the ultimate goal of community policing: a close and cooperative relation between citizen and officer that could enable collective problem solving.

This desire for a responsive police force was tempered for many citizens by the recognition that officers cannot always do precisely what residents request. Citizens frequently indicated that they understood that subservience needed impediments. They accepted the need for officers to uphold the dictates of the law, and to assert their unique authority to calm chaotic and dangerous situations. However, for many, this impulse toward authority was too often overemphasized, too often employed unnecessarily. Citizens often described instances where they felt dismissed, where the police's assertion of superiority left them diminished. When the police assumed such an authoritative stance, the citizenry found themselves unable to influence what officers did. They saw a limit on their ability to make the police subservient to citizen input, and they were frustrated.

They were also often frustrated with the police's construction of community. On the one hand, many residents understood the need for bureaucratic routines, and many shared the police's moral construction of crime and the valiant role of officers in combating it. On the other hand, residents complained about how their requests for service could be distorted, ignored, or routed through an interminable labyrinth of process. Many also argued that, in the police's moral universe, their neighborhood was found wanting, and thus undeserving of serious police attention. Hopes for a more subservient police force were thwarted by police procedure and moralizing.

In short, the projects of subservience, separation, and generativity stand in uneasy relation with one another and are difficult to reconcile in theory and, especially, in practice. The intractability of this tension helps us to understand just why "it is so difficult" to find a way by which community and police, as one important instance of the state-society relation, can move forward toward the goal of collective and cooperative problem solving.

Whether a meaningful way forward exists at all remains an unanswered question. I use chapter 6 to review my analysis, with the aim of assessing the larger lessons it offers and the possibilities it implies for how to best to structure the police-community relation.

The Unbearable Lightness of Community

It is understandable that two leading police scholars would argue that community policing "is the most important development in policing in the past quarter century."[1] No self-respecting police department in the United States—and now much of the world—does *not* have a program that can plausibly be placed under the community policing umbrella. Today, there is little talk of the professional model, little talk of competing models for organizing police departments. Community policing is the preferred ideal.

And why not? Given the desired ends of community policing—more cohesive and politically competent urban neighborhoods, closer ties between residents and officers, more creative police problem-solving—how could one object to its hegemony as a model for police organizations? Why not a more capable and engaged citizenry, why not a more responsive and effective police force?

In preceding chapters, I have addressed these and related questions. I pursued two broad goals. One was to disinter the normative assumptions about community, and about the community-police relation, that underlie legitimations of community policing. I sought to expose and contrast the architecture of differing normative visions of community and of the state-society relation. I both stressed the force of such visions and demonstrated that no single one of them commands absolute allegiance. Such normative competition is inescapable. Any broad evaluation of community policing must acknowledge this competition and the unavoidable political tensions it generates.

But normative theorizing is insufficient to the task of evaluating community policing fully. My second goal was empirical assessment. I engaged two groups of actors, urban citizens and urban police officers, because they are the people meant to create community policing. One can isolate and counterpose normative visions, but such an abstract exercise fails to assess the tractability of any such vision in daily experience. To determine whether the ideals of community policing can be achieved involves collecting and analyzing the qualitative data I relied upon here.

My analysis teaches many lessons. In chapter 1, for instance, I showed that the dominant ideologies of community that captivate normative theorists do not, in fact, have much relevance for urban residents. Certainly, the "thick" visions of communal togetherness that communitarians advocate—either through the recovery of shared values or through the discovery of political possibility—do not resonate deeply for citizens. Yet people want something more than a fleeting, thin connection. Instead, they want relations of basic familiarity upon which they can build a sense of security in place; they want to live in a neighborhood of known others whose actions they can predict and on whom they can rely. The hegemony of this vision of neighborhood connection has implications for our expectations of community political action. If theorists of community politics anchor their prescriptions on the necessity or desirability of close neighborly connections, they will encounter only frustration. Community cannot bear the weight of such expectations.

But perhaps the establishment of relations that breed security is a sufficient basis for the communal political work that community policing envisions. In chapter 2, I explored that possibility. I showed that the residents were quite pessimistic that this could transpire. They consistently cited numerous obstacles that impede their neighborhood's ability to coalesce and to advocate. Activists especially, but others as well, noted several factors that prevented the realization of communal political agency. Importantly, the significance of these obstacles varies unevenly with geography, such that economically disadvantaged communities face larger hurdles to political efficacy than do advantaged ones. Taken together, the analyses of chapters 1 and 2 suggest a need to question endorsements of community policing that presume that neighborhoods can and will enact an efficacious and representative politics.

Further, even if communities could organize effectively, they would still need to encounter a state receptive to their suggestions. Any assessment of community policing thus must determine whether the police, as one symbolically significant arm of the state, demonstrate such receptivity. In chapter 3 I explained how the democratic aspiration for a state that is responsive to its citizenry is not the only legitimate standard for the state-society relation, not the only desirable articulation of the police-community connection. Instead, states might sometimes be best understood as importantly separate from the citizenry, through adherence to either an abstract legal code or a set of professional norms. Also, the state can accurately be described as generative of society, because of its policies, its bureaucratic routines, and its moralized discourses. I used the ethnographic data to demonstrate that all three modes of state-society relation inform the police's approach to the communities they serve; in both theory and practice, these modes are in tension. This reality complicates any blithe endorsement of community policing that presumes a police force that is largely responsive to communal politicking.

Although all three modes are significant to police practice, the narrative of separation has unusual power. In chapter 4, I explained how the police reinforce a self-construction as authoritative crime fighters whose autonomy and discretionary authority deserve respect. Powerful impulses in the culture create a wariness of citizen oversight and an endorsement of officer independence. These realities should also cause us to reconsider any high expectations we might have for community policing; officers continue to resist the suggestion that they make themselves especially subservient to community input.

This helps explain much resident frustration with the police—a phenomenon I explored in depth in chapter 5. I showed that the three modes of state-society relation all possess legitimacy in the eyes of the citizenry. This suggests that possible tensions between these modes irrevocably complicate the politics of police-community relations. Yet residents do appear to recognize the comparative power of the narrative of separation, and they express dissatisfaction with it. They also sometimes chafe at the routines the police use for both apprehending their input and morally constructing their neighborhoods. The frequency of these complaints suggests that much work must be done before police-community relations can improve significantly.

In short, community is unbearably light: it cannot bear the political weight of projects like community policing, its voice is not loudly heard by state agencies like the police. In this chapter I explore the implications of this conclusion, by addressing four sets of questions. The first concerns whether the hegemony of community policing deserves to persist. To conclude that community is unbearably light is to suggest that it is time to reassess our expectations of police-community relations, to reconsider the status of community policing as a model. The second focuses on the broader role of the state in community. As agents of coercive force, the police should not be a prominent player in building community. If the state wishes to help create effective communal organizations, then it must avoid over-reliance on the police to help accomplish this.

Of course, there is no simple pathway for more legitimate state-society relations. My third set of questions revolves around the tensions that ensue from the normative competition between subservience, separation, and generativity. The intractability of these tensions means that they deserve our continuing attention, not because any easy resolution is possible, but because normative choices simply must be made. They are best made consciously, and with full awareness of the consequences that inevitably attend any such choice.

My final set of questions concerns what role, if any, community should play in urban politics. Because community as an ideal will not disappear, it is senseless to discard it as a potential vehicle through which localized politics can emerge. Yet skepticism about its utility must be ever present; its political lightness must never escape our attention.

On the Status of Community Policing

Skogan and Roth accurately describe the rapture that surrounds community policing. Its popularity allowed it to overthrow the professional model as the organizational ideal, and it infected much popular and political discourse about the proper role of police departments. Yet mine is not the first analysis to suggest that the reality of community policing does not match its rhetoric. Police officers continue to resist it with appreciable tenacity.[2] Many departments appear to adopt community policing programs not because of a strong internally held belief in its utility, but primarily to maintain their legitimacy, particularly in the eyes of outside funding agencies like the federal government.[3] Even where community policing does exist to a noticeable extent, it is typically confined to a single, often isolated unit within a department, rather than diffused throughout the organization as a culture-changing reform.[4] In addition, the police's community engagement is often largely confined to working with advantaged neighborhoods whose

perspectives on crime and policing are congruent with those of the officers.[5] More marginalized groups rarely get a word in edgewise.

Indeed, we do not possess much evidence that the police-community
relation has changed much as a consequence of community policing.
Take, for example, the case of Chicago, whose municipal government
undertook what is arguably the most ambitious effort yet to implement
community policing. Chicago's city government spearheaded an
impressively concerted effort at a thoroughgoing reform. This included
making numerous city government agencies responsive to community
concerns about a range of issues connected to crime and disorder.[6] It
also included a robust effort to advertise and convene regular police-
community forums. In their analysis of these forums, Wesley Skogan
and his associates found that awareness of community policing in
Chicago increased measurably, and that about 28 percent of Chicagoans
who were aware of police-community forums had attended at least one.
Yet the number of attendees did not increase much during the late
1990s, and it consisted mostly of people who showed up regularly. More
importantly, observations of these meetings noted a continuing difficulty in establishing interactions in which both sides participated
actively to create strategies to solve problems and to establish means to
evaluate those strategies.[7] Skogan et al. conclude that these forums have
not "become a general vehicle for the kind of systematic problem solving that the department envisions."[8] Similarly, Roth's extensive evaluation of community policing programs in the United States leads him to
argue that "True community partnerships, involving sharing power and
decision making, are rare at this time."[9]

But why should we expect anything different? Why did we ever
believe that communities could be more capable of articulating a political voice? Why did we ever believe that the police would be especially
responsive?[10] In retrospect, these assumptions seem naive.[11] They may
even be dangerous. To hold onto the hope that these dynamics will
change is politically untenable.

Such hope, however well-intentioned, can lead us astray. For
instance, its perpetuation may cause us to neglect the significant challenges that face neighborhoods of economic disadvantage. As Ralph
Taylor persuasively argues, broader factors—the cost of housing, the
availability of jobs, the state of various social services—all deeply impact
the everyday reality of urban neighborhoods far more than factors of
"disorder."[12] And none of these broader factors are in any way touchable
by the police. Further, as Eric Klinenberg notes,[13] efforts to foreground
policing in urban politics necessarily means that fear of crime receives

significant attention. This can mean that discussions about how to improve broader measures of social welfare get sidetracked. It can also increase levels of distrust and suspicion within the urban citizenry, as the fearful seek protection from the fear-inducing. In Klinenberg's strong words, "As a program for civic renewal, [community] policing represents democracy in its most desperate and depraved form."[14]

Indeed, if community regeneration is a priority, why foreground the police in any such effort? As agents of coercive force, the police do not make especially good community builders. As I explained in chapter 4, this coercive capacity deeply shapes police culture in ways that reinforce officers' sense of separation from community: they are the authoritative crime fighters.[15] Further, their dominant narrative for crime causation—as the result of ill-minded bad apples—means that officers rarely pay attention to wider community dynamics and may in fact exacerbate divisions within communities.[16] Police officers may be able to make urban residents feel safer, as when they patrol more visibly on foot.[17] But this hardly translates into any competence in restoring the deeper fabric of community life. Indeed, given the police's standoffishness toward community politics, it seems positively counterproductive to expect neighborhood political regeneration through community policing.[18]

This misplaced faith in community policing may help reduce police accountability. To presume that police-community forums will work to ensure citizen oversight of the police is a mistake; the evidence suggests otherwise. A more productive strategy would be to bolster formal mechanisms of civilian oversight, such as civilian review boards. These may be a more effective means to ensure that officers are held accountable to community expectations, as expressed through formal proscriptions on police behavior. The police successfully resist the idea that community forums are places where they can be called to task. Accountability must be pursued through alternate means.

In short, it is better to constrict, not expand, the role of the police in any project aimed at community regeneration. The police's coercive role means that they are not good community builders, and thus we fool ourselves—and harm communities—by holding onto the more elaborate rhetoric of community policing. As much as citizens themselves desire a friendly relation with the cop on the beat, this should not obscure the impressive array of coercive tools that officers carry, and that so fundamentally structure their culture and job orientation. These coercive mechanisms are deeply implicated in other developments that occurred concurrently with the ascent of community policing—including, most notably, the growth of paramilitary policing and the rising popularity of broken windows policing, neither of which exemplifies a friendly police force.[19] Indeed, the era of community policing has seen a robust reinforcement of the "get tough" rhetoric of crime control, and the concomitant policy changes that make the United States the world leader in incarceration.[20] Further, these punitive practices reinforce the discrepancies between urban neighborhoods, because they impact poor and minority-dominated neighborhoods most significantly. Residents of such neighborhoods are more likely to be incarcerated, and thereby to suffer the long-term consequences of the stigma of conviction; both steady employment and family stability become more elusive with a history of incarceration.[21] This reality merely increases the level of distress facing urban neighborhoods of disadvantage.[22]

This "get tough" orientation obviously has much resonance within the culture of the police, which helps explain officers' inability to work productively in creating equitable partnerships with community groups. The professional model lives on, quietly flourishing beneath the reform rhetoric of community policing.

As coercive agents, the police do have a role to play. They remain a vital means of resolving immediate crises, of promoting a sense of safety, of responding to specific instances of criminality. They can even competently work to prevent crime, though such efforts have historically been inconsistent.[23] But to entrust them with these responsibilities hardly means that we should expect anything more, that we should hope for them to be rendered easily subservient to community oversight. As Berkley long ago recognized, "The phrase 'democratic police force' is a contradiction in terms."[24]

In short, perhaps it is time to abandon community policing, to give up the ghost of a chance of its ever representing a project for meaningful police reform. Its promise remains unrealized, largely because it was never realizable. This does not mean that police departments should discontinue efforts to forge productive working relations with community groups. What Loader and Mulcahy note for England holds true for the United States: that the large reservoir of public support for the police provides a solid basis upon which constructive police-community relations can be built.[25] But such constructive relations are frustrated when the police act as if isolated programs constitute reform or as if they presently engage in meaningful power sharing with community groups. Given the limits to what the police can accomplish, they should withdraw from a prominent place in efforts to revive urban neighborhoods, a role that more properly belongs to other agencies of the state.

The State and Community

The expected role of the police highlights a paradox. They are a state agency expected to reduce crime and to rebuild community, yet these goals are rendered well nigh unattainable by other components of the state. The police are not capable of altering the social dynamics that most affect urban neighborhoods. These dynamics, however, are implicated in policies enacted by the state. Indeed, Robert Sampson and Janet Lauritsen show that federal policies from urban renewal to public housing may have done more to cause inner-city violence than to prevent it.[26] At other levels of government as well, policies controlling zoning, housing, economic development strategies, and education strongly shape the realities that confront urban communities.[27] It is hardly the police's fault that they cannot do much to stem the ill effects of the concentrated disadvantage these policies enable. Even if greater competence in police problem-solving operations is possible, it would be but a puny bulwark against the overwhelming challenges generated by broader forces. The work of the police, in effect, represents an attempt by one component of the state apparatus to minimize the negative effects generated by other components.

This reality helps underwrite a necessary skepticism about the possibilities of community policing and mandates a shift of our attention toward those state activities that more directly bear on urban neighborhoods. It is unrealistic to expect the police to undo the catastrophic consequences of state policies that help generate urban landscapes of widely variant economic and political possibility. To create and reiterate such expectations for the police leads to an unjustifiable neglect of the role of other arms of the state in rendering those expectations impossible to meet. Recall the words of Marshall, the pastor in Centralia quoted in the last chapter:

I mean hey, [the police] keep me safe at night, I appreciate them for doing that. If they want to focus on that and cleaning up the streets, I say you concentrate on that and let us, and find other agencies or whatever, or have the police come alongside churches that can help change the community. But you put the police in charge of changing the community as well as policing the community, they are going to fail in both areas.[28]

This is a point very well taken. The police *cannot* change the underlying conditions that shape urban neighborhoods in any substantive fashion. Whatever culpability the state possesses for the health of community largely lies outside the police's rightful charge. Let us absolve the police of unjustified responsibilities, and hold the state accountable for those of its policies most directly connected to the realities that urban neighborhoods confront daily.[29]

Let us also hold the state accountable for its responsibility in generating the means by which citizen agency can be expressed and reinforced. The insight that the state largely generates the community with which it interacts requires us to assess just how this transpires. The state and civil society, in other words, are not strictly separate and opposing entities, but importantly connected by the opportunities the state constructs for citizen activity and involvement.[30] Through various means—wars, schools, political parties—states have historically created the mechanisms for citizen participation in public life. Without such state-sponsored supports, citizen action is unlikely to build. As Ehrenberg summarizes it, "You cannot explain civil society apart from the influence of state-building."[31] Sadly, as I have demonstrated, community policing does not represent an instance of the state's nurturing of civil society. Yet this failure should not spell the end of the search for alternate means by which this can transpire. Because the police are hobbled by their coercive power from being supports for a revived civic politic, the state must find other means to accomplish this.[32]

On the State-Society Relation

I have been at pains to establish that there is no simple means by which the state-society relation should be pursued. That is because different modes of that relation possess legitimacy. These modes can coexist peacefully. Certainly the protection of individual and group liberties is essential for any modicum of democratic activity; subservience and separation are therefore necessarily twinned.[33] Yet conflict between these modes is unavoidable. In the case of the police, the drive to hold officers subservient to citizen input is frequently thwarted by the police's ability to remain separate and by their generative practices. Importantly, the police maintain their separation more through their adherence to internally defined norms of professional conduct than through their loyalty to the legal code. Whatever merit attends to professional authority, its assertion can reduce the extent of citizen oversight. Similarly, the frames the police use to structure citizen input often yield citizen frustration, a commonly held belief that residents cannot make themselves heard as they desire.

There is no magical resolution to the dilemma posed by conflicting paths to legitimate state-society relations. Yet this does not mean that this dilemma should escape our continued attention. In fact, it is precisely the lack of a resolution that mandates such attention. Certainly, no single path should capture our unalloyed allegiance, because each possesses shortcomings. The single-minded pursuit of subservience falls prey to the well-grounded liberal complaint about the potential excesses of majority rule. Similarly, a state agency that takes refuge behind separation may find its legitimacy challenged. Resolute adherence to the legal code may frustrate citizen groups that expect a more flexible response; the preservation of professional authority can unjustifiably stiff-arm citizen oversight. In addition, the ways in which the state generates community also deserve perpetual examination. State policies that deeply affect urban neighborhoods need to be measured against ideals of justice, fairness, and equity, especially if we expect localized groups to assume responsibility for asserting a political voice. The practices by which state agencies apprehend those citizen voices, while unavoidable, can be overly restrictive filters. And the moral discourse that state actors invoke, also unavoidable, can similarly mute citizen input.

At the same time, we cannot do without any of these articulations of the state-society relation. The norm of democracy is so well-established that some evidence of its perpetuation is necessary for state legitimacy; citizens rightly expect a responsive state. Similarly, they rightly expect the state to be checked in its exercise of power. State actors must be accountable to legal rules and to rights-bearing citizens. And state actors will continue to construct and defend professional norms for regulating their conduct. Because such norms can improve efficiency and performance, they are not necessarily problematic. Further, the state cannot help but generate community. It will continue to construct policies that deeply impact urban neighborhoods, to use bureaucratic routines to apprehend citizen input, and to develop moral justifications for its actions.

It is therefore unproductive to suggest that any of these three modes of state-society relation deserves absolute priority over another. Our political attention should focus instead on the particular expression of these modes and the balance between them. We can sensibly query whether state agencies like the police do or do not respond to citizen suggestions, whether they are or are not constrained by legal regulations. We can assess their professional practices, both to determine their effectiveness and to assess whether their expression illegitimately constricts public oversight. We can ask whether an overly robust moralism allows state agencies to elevate their actions beyond the pale of assessment.[34] Further, we can wonder whether the state's moral desire to protect its citizenry should not extend beyond the crusade-like quest to apprehend and incarcerate criminal "predators." A moral case can be easily made to suggest that the state, as well, should ensure basic levels of economic security.

Indeed, such a commitment to the less advantaged is under sustained attack in these neo-liberal times. With the economic market the increasingly hegemonic metaphor for structuring social and political relations, citizens are seen primarily as free economic agents, expected to ensure their own survival with minimal support. If they do approach the state, they are more likely to be viewed as clients than as citizens; perhaps they will receive limited services, but their political agency will likely not be recognized. The pervasiveness of market-based individualism stands as an obvious contrast to nostalgic visions of communal togetherness and political capability, and it is a glaring contradiction of neo-liberal discourse that it often trumpets both simultaneously.[35] My larger point is that the state's moralism is an inescapable fact, regardless of the ideal of liberalism that the state remain as neutral as possible. Yet the form of the state's moral framework remains susceptible to change, even though it is often situated on a transcendental plane ostensibly beyond critique. The state's moral framework, and the power accorded it, deserve perpetual questioning.

It would be simpler, of course, to find some Archimedean point upon which one could situate the ideal state-society connection. Yet such a quest is fruitless. There will *always* be tension between these modes of relation, *always* be a charged politics about their content and their competition. Our obligation is less to locate an ideal end state than to monitor the balance continually and strive toward some basic degree of equilibrium.

In the case of the police, the politics of the state-society connection will be especially fervent, because of the coercive power officers possess. At present, the evidence suggests that the balance is not ideal, that the police lie rather too far beyond the reach of citizen oversight. Further, community policing is a poor vehicle through which a better balance can be struck. Because police so resist informal oversight, they need to be held accountable through more formal bodies. Citizens need sufficient means by which they can raise questions about police operations. Although the police should be granted some deference, they must also confront clear limits on their authority. These limits must be imposed by legal regulations and enforced by independent bodies. And police morality must be checked to ensure that it does not legitimate questionable actions. Indeed, the police's capacity to construct crime in highly moralized language, and thereby to cast their own actions as unassailable, deserves ongoing suspicion. Otherwise, officers can arrogate to themselves a degree of power that is unacceptable in ostensibly democratic societies.

In this ongoing political struggle over the state-society connection, it is questionable whether "community" should play a central role, to which issue I now turn.

On the Political Status of Community

It is easy to understand the persistent longing for community. That is because community promises so much—togetherness, support, cooperation, care, compassion, friendship. No social institution can provide these in quite the same way. Certainly, the state cannot do it.[36] In an increasingly mobile and tenuous world, the allure of a secure anchor is irresistible. In her extensive interviews with volunteer activists, Nina Eliasoph witnessed this allure: "Almost everyone I met harbored intense nostalgia for the warm, totally enveloping community. They all ambivalently recognized that without anything thicker than moral minimalism holding anyone together, life could be lonely and dangerous."[37] The promise of localized democracy is similarly seductive; grass-roots empowerment is easy to desire and to support. It is thus not surprising that programs like community policing galvanize such strong enthusiasm from across the political spectrum.

These hopes for community are sometimes realized. Projects done in the name of community can work to improve the life circumstances of urbanites, even those in neighborhoods of disadvantage.[38] Yet assessments of community politics suggest that these successes are not typical, that urban neighborhoods only rarely approximate the social and political ideals many hold out for community.[39] Certainly, none of the neighborhoods I studied in West Seattle do, for the reasons I delineated. Because these realities exist across American cities, my cautionary tale about the unrealized hopes for community deserves consideration when approaching projects like community policing.

In fact, the seductiveness of localized democracy should not blind us to its very real downsides. The devolution of power toward smaller units can work to increase inequality, both within and across groups. Within groups, there exists always the possibility that the more voluble and energetic members of a small faction can dominate the discussion and push an agenda that is not widely supported.[40] Across groups, certain community organizations typically possess greater political skills than others and can thereby prosper when there is competition between groups for government resources. Because neighborhoods of higher economic standing are more likely to harbor members with political acumen, devolution of political authority can thereby increase the social and economic gaps between rich areas and poor ones.[41] It can also decrease the possibilities for political alliances across social groups. If groups are perpetually in competition with one another, they may well find it difficult to forge coalitions.[42]

The devolution of authority can also frame problems as local when they are actually the result of dynamics generated at other scales.[43] Because urban neighborhoods are shaped by regional, national, and international forces, perhaps it is misguided to expect much from local political organizations, regardless of their representativeness or seeming capability.[44] If this is true, citizens might rightfully focus their political energies on national-level political groups, even if that means that they are increasingly "bowling alone."[45] Certainly, neighborhood political organizations that fail to connect their agendas to larger political dynamics stand little chance of seeing meaningful long-term change.[46] One can therefore call into question the presumption that localized groups should take such extensive responsibility for ensuring their own economic and social security. Even the best organized of such groups can do little to affect broader dynamics; communities of disadvantage face nearly impossible odds.[47]

This concern about increasing the degree of disadvantage is particularly regnant in these post-welfare, neo-liberal times. The extent of concentrated disadvantage in American cities remains exceedingly high, the political will to create a sense of collective interest exceedingly weak. The ideology of devolved authority can certainly sound sweet; who, after all, wants to discourage local empowerment? But if localized politics obscure larger-scale dynamics, if they render more difficult the creation of a public interest, then they deserve questioning.[48]

This is not to suggest that neighborhoods cannot and should not be the focus of political organizing. Because the problems that confront neighborhoods are immediate, they can often be a visible and direct reminder to citizens of the importance of political action.[49] In the case of crime, neighborhoods can work as a locus for the development of social capital and informal social control, both of which can help reduce crime and collective fear.[50] But, as Clarence Stone and his associates note, social capital is different from civic capacity.[51] In other words, the development of social connections to provide support and reciprocal assistance is not the same as the capacity to act effectively in the wider body politic.

In short, simply to devolve authority to "the community" without cognizance of the immense obstacles many urban neighborhoods face in gaining greater economic, social, and political equality confounds justice rather than promotes it.

The Unbearable Lightness of Community

Now here, I've known of three drive-by shootings and I know all the kids that have been shot. That creates an awareness. But what can community really do to fix that, I'm not sure.
MARSHALL, CENTRALIA RESIDENT

Little that I learned in West Seattle offers much reason for optimism about the political power of community. Neighborhood groups rarely organize effectively, focus broadly, or attract much state attention. Although the pursuit of communal connections will not abate, these connections cannot typically bear much political weight. Our hopes for community as a means for generating political energy must always remain modest.

As Marshall notes, community members possess minimal capacity to alter the dynamics of violent crime. This is especially the case if such violence can be traced ultimately to matters of economic and racial disadvantage. Community can, of course, exist as a central component of many people's lives; urban citizens continue to seek connections of meaning, value, and support, both within and outside their neighborhoods. In many cases, these neighborhoods can possess a skein of relations capable of exerting informal social control sufficient to help reduce rates of crime. Certainly, as a core sociological and political concept, community is unlikely to disappear anytime soon.

Yet we invest it with significant political hopes at great potential peril. It would certainly be nice if one could conclude otherwise, if more romanticized notions of community togetherness and empowerment actually had more resonance and possibility. A more clear-eyed assessment points us in the opposite direction, toward a necessary skepticism about the promise of community policing and other efforts to increase localized self-governance. Such an assessment makes clear that community cannot be expected to be a robust force for political action.

It also reveals that community, in whatever form, cannot relate in any simple fashion to the police or any other component of the state. If the goals are to increase the political capacity of urban neighborhoods and to improve citizen oversight of state agencies like the police, community is unlikely to be an effective vehicle. For now, and for the foreseeable future, it is best to recognize that, all too often, community is unbearably light.

Notes

INTRODUCTION

1. Colin Bell and Howard Newby, *Community Studies: An Introduction to the Sociology of the Local Community* (New York: Praeger, 1974), 15. Robert Sampson makes much the same point: "Community seems to be the modern elixir for much of what ails American society." Sampson, "What 'Community' Supplies," in *Urban Problems and Community Development*, ed. Ronald Ferguson and William Dickens (Washington, D.C.: Brookings Institution Press, 1999), 241–92.
2. Michael Sandel, *Liberalism and the Limits of Justice* (Cambridge: Cambridge University Press, 1998), 183.
3. Adam Crawford overstates the case when he says that, in daily use, community policing possesses "unfathomable plasticity." Crawford, *The Local Governance of Crime* (Oxford: Clarendon Press, 1997), 45. Yet this is a reminder that the term is broad. On the various ways community policing has been implemented in the United States, see Edward Maguire and Stephen Mastrofski, "Patterns of Community Policing in the United States," *Police Quarterly* 3 (2000): 4–45.
4. That Weed and Seed does not work to provide opportunities for extensive citizen oversight is documented well in Lisa Miller, *The Politics of Community Crime Prevention* (Aldershot: Ashgate, 2001). Miller's analysis focuses on two Weed and Seed operations in Seattle.
5. Stephen Holmes makes this point provocatively: "When we hear [community], all our critical faculties are meant to fall asleep." Holmes, "The Permanent Structure of Antiliberal Thought," in Nancy Rosenblum (ed.), *Liberalism and the Moral Life* (Cambridge: Harvard University Press, 1989), 230.

6. As with all names in this book, this is a pseudonym.

7. Not only is West Seattle a place where community might be a potent source of social and political connection; Seattle itself possesses a strong history of grassroots political activity and activism. On this history, see: Margaret Gordon, Hubert Locke, Laurie McCutcheon, and William Stafford, "Seattle: Grassroots Politics Shaping the Environment," in *Big City Politics in Transition,* ed. Hank Tavitch and John Thomas (Newbury Park, Calif.: Sage Publications, 1991); Edward Banfield, *Big City Politics* (New York: Random House, 1965); Andrew Gordon, Hubert Locke, and Cy Ulberg, "Ethnic Diversity in Southeast Seattle," *Cityscape: A Journal of Policy Development and Research* 4 (1998): 197–219. Seattle also once occupied a spot among the leading practitioners of community policing. See Dan Fleissner, Nicholas Fedan, Ezra Stotland, and David Klinger, *Community Policing in Seattle: A Descriptive Study of the South Seattle Crime Reduction Project* (Seattle: Seattle Police Department, 1991). For these reasons, Seattle ostensibly stands positioned to enact a successful effort at community policing.

8. Views to the west capture the Olympic Mountains, to the north downtown, and to the east the Cascade Mountains. Especially well-situated properties capture all three.

9. Demographic data come from the 2000 United States Census. Data on crime and calls for police service come from the Seattle Police Department.

10. At the time of the research, Blufftop was undergoing a major reconstruction as part of a federally sponsored "Hope VI" project. The goal was to create new structures that would ultimately include a mix of owner-occupied homes and federally subsidized rentals. This transformation included moving families out of the facility during the reconstruction. These families were either placed in alternate public housing accommodations or given vouchers to enter the private housing market. Some of them will return, although many, if not most, will not.

11. There was a change in leadership of the precinct during the period of the research.

12. Such an evaluative task is undertaken elsewhere. See Tim Hope, "Community Crime Prevention," in *Building a Safer Society,* ed. Michael Tonry and David Farrington (Chicago: University of Chicago Press, 1995); Lawrence Sherman, "Communities and Crime," in *Preventing Crime: What Works, What Doesn't, What's Promising,* ed. Lawrence Sherman (Washington, D.C.: U.S. Department of Justice, 1997); Wesley Skogan and Susan Hartnett, *Community Policing, Chicago Style* (New York: Oxford University Press, 1997); Wesley Skogan, ed., *Community Policing: Can It Work?* (Belmont, Calif.: Wadsworth/Thomson, 2004).

13. See, for instance: Randolph Grinc, "Angels in Marble: Problems in Stimulating Community Involvement in Community Policing," *Crime & Delinquency* 40 (1994): 437–68; Crawford, *Local Governance of Crime*; William Lyons, *The Politics of Community Policing: Rearranging the Power to Punish*

(Ann Arbor: University of Michigan Press, 1999); Miller, *The Politics of Community Crime Prevention*; Wilson Edward Reed, *The Politics of Community Policing: The Case of Seattle* (New York: Garland, 1999); Susan Sadd and Randolph Grinc, "Innovative Neighborhood Oriented Policing: An Evaluation of Community Policing Programs in Eight Cities," in *The Challenges of Community Policing: Testing the Promises,* ed. Dennis Rosenbaum (Thousand Oaks, Calif.: Sage Publications, 1994), 27–52; Ralph Saunders, "You Be Our Eyes and Ears: Doing Community Policing in Dorchester," Ph.D. dissertation, Department of Geography, University of Arizona, 1997; Brian Williams, *Citizen Perspectives on Community Policing: A Case Study in Athens, Georgia* (Albany: SUNY Press, 1998).

14. Evidence suggests that similar patterns emerge in other countries. For analysis of Great Britain, see Trevor Bennett, "Community Policing on the Ground: Developments in Britain," in *The Challenges of Community Policing: Testing the Promises,* ed. Dennis Rosenbaum (Thousand Oaks, Calif.: Sage Publications, 1994); for Canada, see Benedikt Fischer and Blake Poland, "Exclusion, Risk and Governance: Reflections on 'Community Policing' and 'Public Health,'" *Geoforum* 29 (1998): 187–97; for Kenya, see Mutuma Ruteere and Marie-Emmanuelle Pommerolle, "Democratizing Security or Decentralizing Repression: The Ambiguities of Community Policing in Kenya," *African Affairs* 102 (2003): 587–604.

15. Seattle has also been the focus of much recent scholarly attention on its efforts at policing reform. See Lyons, *The Politics of Community Policing*; Miller, *The Politics of Community Crime Prevention*; Reed, *The Politics of Community Policing*. In broad strokes, my analysis reinforces these prior works, especially the finding that the police are less than fully responsive to the wide range of communal input they receive. My analysis differs in two ways: (1) it builds a more robust theoretical framework for understanding both communal political action and the nature of the state-society relation; and (2) it relies upon a somewhat broader data base, and thus provides a deeper empirical basis for evaluating community policing in Seattle.

16. See Steve Herbert, "For Ethnography," *Progress in Human Geography* 24 (2000): 550–68.

17. See Claude Fischer, Robert Max Jackson, C. Ann Stueve, Kathleen Gerson, Lynne McCallister Jones, with Mark Baldassare, *Networks and Places: Social Relations in the Urban Setting* (New York: The Free Press, 1977); Barry Wellman and Barry Leighton, "Networks, Neighborhoods, and Communities: Approaches to the Study of the Community Question," *Urban Affairs Quarterly* 14 (1979): 363–90; Melvin Webber, "Order in Diversity: Community without Propinquity," in *Cities and Space: The Future Use of Urban Land,* ed. Lowdon Wingo (Baltimore, Md.: Johns Hopkins University Press, 1963).

18. Claude Fischer, *To Dwell among Friends: Personal Networks in Town and City* (Chicago: University of Chicago Press, 1982).

19. Margit Mayer, "Urban Movements and Urban Theory," in *The Urban Moment: Cosmopolitan Essays on the Late-20th-Century City,* ed. Robert Beauregard and Sophie Body-Gendrot (Thousand Oaks, Calif.: Sage Publications, 1999), 209–38; Mark R. Warren, *Dry Bones Rattling: Community Building to Revitalize Democracy* (Princeton, N.J.: Princeton University Press, 2001). William Sites argues that neighborhood-based activism frequently founders precisely because of the inability to forge wider connections. See Sites, *Remaking New York: Primitive Globalization and the Politics of Urban Community* (Minneapolis: University of Minnesota Press, 2003).

20. Anne Phillips, "Feminism and the Attractions of the Local," in *Rethinking Local Democracy,* ed. Desmond King and Gerry Stoker (London: Macmillan, 1996).

21. See Avery Guest, Barrett Lee, and Lynn Staeheli, "Changing Locality Identification in the Metropolis," *American Sociological Review* 47 (1982): 543–49; Albert Hunter, *Symbolic Communities: The Persistence and Change of Chicago's Local Communities* (Chicago: University of Chicago Press, 1974); Lyon, *The Community in Urban Society*; Gerald Suttles, *The Social Construction of Communities* (Chicago: University of Chicago Press, 1972). The centrality of localized ties for projects of community governance focused on crime is elaborated in Todd Clear and David Karp, *The Community Justice Ideal: Preventing Crime and Achieving Justice* (Boulder, Colo.: Westview Press, 1999).

22. Albert Hunter, "The Urban Neighborhood: Its Analytical and Social Contexts," *Urban Affairs Quarterly* 14 (1978): 267–88, at 285.

23. Sampson, "What `Community' Supplies," 247.

24. As Chaskin and his coauthors put it: "In common parlance, the term *community* is often used interchangeably with *neighborhood* to refer to a geographic area within which there is a set of shared interests or symbolic attributes. In the field of community building . . . policymakers and practitioners assume that sufficient commonality of circumstance and identity exists within the geographic boundaries of neighborhoods to develop them further as 'communities.' " Robert Chaskin, Prudence Brown, Sudhir Venkatesh, and Avis Vidal, *Building Community Capacity* (New York: Aldine de Gruyter, 2001), 8. This, of course, does not mean that political projects done in the name of "community" do not transcend neighborhood boundaries. Indeed, the evidence suggests that the success of such projects depends precisely upon this translocal orientation. See, for instance, Warren, *Dry Bones Rattling*.

CHAPTER ONE

1. Quoted in Thomas Bender, *Community and Social Change* (New Brunswick, N.J.: Rutgers University Press, 1978), 6.

2. A useful review of this extensive literature is found in Larry Lyon, *The Community in Urban Society* (Philadelphia: Temple University Press, 1987).

3. Benjamin Barber, *Strong Democracy: Participatory Politics for a New Age* (Berkeley and Los Angeles: University of California Press, 1984).

4. Ibid., xxiii.

5. If disagreements within communal groups are to be resolved respectfully, the need for deliberation becomes obvious. The utility and procedures for this deliberation thus command much attention from theorists who seek a democratic project that involves engaged citizens who can accommodate differences of opinion. See, for instance: Susan Bickford, *The Dissonance of Democracy: Listening, Conflict and Citizenship.* (Ithaca, N.Y.: Cornell University Press, 1996); James Bohman, *Public Deliberation: Pluralism, Complexity and Democracy* (Cambridge, Mass.: MIT Press, 2000); Simone Chambers, *Reasonable Democracy: Jurgen Habermas and the Politics of Discourse* (Ithaca, N.Y.: Cornell University Press, 1996); Amy Gutman and Dennis Thompson, *Democracy and Disagreement* (Cambridge: Harvard University Press, 1996); Jurgen Habermas, *The Theory of Communicative Action, Reason and Rationalization of Society, Volume I* (Boston: Beacon Press, 1984).

6. The assertion of the centrality of politics in the life of community is a disputed one, a point explored in more depth below. See Stephen Gardbaum, "Law, Politics and the Claims of Community," *Michigan Law Review* 90 (1991): 685–760.

7. Robert Ellickson, "New Institutions for Old Neighborhoods," *Duke Law Journal* 48 (1998): 75–122; Frank Ryan, *Real Democracy: The New England Town Meeting and How It Works* (Chicago: University of Chicago Press, 2004).

8. Jeffrey Berry, "The Rise of Citizen Groups," in *Civic Engagement in American Democracy,* ed. Theda Skocpol and Morris Fiorina (Washington, D.C.: Brookings Institution Press, 1999), 367.

9. Adrian Little, *The Politics of Community: Theory and Practice* (Edinburgh: Edinburgh University Press, 2002).

10. For overviews of the literature on the geography of crime, see Robert Bursik and Harold Grasmick, *Neighborhoods and Crime: The Dimensions of Effective Community Control* (New York: Lexington Books, 1993); Keith Harries, *The Geography of Crime and Justice* (New York: McGraw-Hill, 1974); Albert Reiss and Michael Tonry, eds., *Communities and Crime* (Chicago: University of Chicago Press, 1986); George Rengert, *The Geography of Illegal Drugs* (Boulder, Colo.: Westview Press, 1996).

11. See, most significantly, Robert Sampson, Stephen Raudenbush, and Felton Earls, "Neighborhoods and Violent Crime: A Multilevel Study of Collective Efficacy," *Science* 277 (August 1997): 918–24. See also Stephanie Greenberg and William Rohe, "Informal Social Control and Crime Prevention in Modern Neighborhoods," in *Urban Neighborhoods: Research and Policy,* ed. Ralph Taylor (New York: Praeger, 1986).

12. Elijah Anderson, *Streetwise: Race, Class and Change in an Urban Community* (Chicago: University of Chicago Press, 1990); Sally Engle Merry, *Urban*

Danger: Life in a Neighborhood of Strangers (Philadelphia: Temple University Press, 1981).

13. There is now a large literature on community and criminal justice. Useful overviews are provided by Todd Clear and David Karp, *The Community Justice Ideal: Preventing Crime and Achieving Justice* (Boulder, Colo.: Westview Press, 1999); Adam Crawford, *The Local Governance of Crime* (Oxford: Clarendon Press, 1997).

14. There is also a large literature on community policing. The history and philosophical orientations of the approach are reviewed in a number of key works: Gary Cordner, "Community Policing: Evidence and Effects," in *Community Policing: Contemporary Readings,* ed. Geoffrey Alpert and Alex Piquero (Prospect Heights, Ill.: Waveland Press, 1998); Jack Greene and Stephen Mastroski, eds., *Community Policing: Rhetoric or Reality?* (New York: Praeger, 1988); Mark Harrison Moore, "Problem-Solving and Community Policing," in *Modern Policing,* ed. Michael Tonry and Norval Morris (Chicago: University of Chicago Press, 1992), 99–158; Wesley Skogan and Susan Hartnett, *Community Policing, Chicago Style* (New York: Oxford University Press, 1997); Robert Trojanowicz and Bonnie Bucqueroux, *Community Policing: How to Get Started* (Cincinnati, Ohio: Anderson, 1994).

15. Robert Fogelson, *Big-City Police* (Cambridge: Harvard University Press, 1977); Samuel Walker, *A Critical History of Police Reform* (Lexington, Ky.: Lexington Books, 1977).

16. Robert Fogelson, "White on Black: A Critique of the McCone Commission Report," in *The Politics of Riot Commissions,* ed. Anthony Platt (New York: Macmillan, 1971), 307–34; Homer Hawkins and Richard Thomas, "White Policing of Black Populations: A History of Race and Social Control in America," in *Out of Order? Policing Black People,* ed. Ellis Cashmore and Eugene McLaughlin (London: Routledge, 1991), 65–86.

17. Moore, "Problem-Solving and Community Policing"; Jerome Skolnick and David Bayley, *The New Blue Line: Police Innovation in Six Cities* (New York: Free Press, 1986).

18. Herman Goldstein, *Problem-Oriented Policing* (New York: McGraw-Hill, 1991); Malcolm Sparrow, Mark Moore, and David Kennedy, *Beyond 911: A New Era for Policing* (New York: Basic Books, 1990). It is important to note that community policing and problem-solving policing are distinct enterprises, even if they are often conflated. The latter does not necessarily depend upon significant amounts of community input, but typically gives greater credence to the police's unique capacities to analyze crime patterns. Nick Tilley argues that the distinction between the two needs to be made more firm, because crime dynamics are often influenced by factors that range across different communities, and because citizens are not especially well equipped to understand those dynamics with any sophistication. See Nick Tilley, "Community Policing and Problem Solving," in *Community*

Policing: Can it Work? ed. Wesley Skogan (Belmont, Calif.: Wadsworth/Thomson Publishing, 2004), 165–84.

19. One early review found ninety-four separate uses of the term "community." George Hillery, "Definitions of Community: Areas of Agreement," *Rural Sociology* 20 (1955): 779–91.

20. In isolating these differences between approaches to community, I do not mean to imply that every theorist adopts only one of them. In any categorical scheme like this, some complexity is inevitably washed out. That said, there are clear differences in approaches to community as political actor, and my schema isolates the central components of those differences. See also Richard Schragger, "The Limits of Localism," *Michigan Law Review* 100 (2001): 371–472.

21. Robert Putnam, *Bowling Alone: The Collapse and Revival of American Community* (New York: Simon and Schuster, 2000).

22. Robert Ellickson, *Order Without Law: How Neighbors Settle Disputes* (Cambridge: Harvard University Press, 1991).

23. Michael Sandel, *Liberalism and the Limits of Justice* (Cambridge: Cambridge University Press, 1998). See also Alistair MacIntyre, *After Virtue: A Study in Moral Theory* (London: Duckworth, 1981); Charles Taylor, *Sources of the Self: The Making of Modern Identity* (Cambridge: Harvard University Press, 1989). For a useful overview of this communitarian line of reasoning, see Stephen Mulhall and Adam Swift, *Liberalism and Communitarianism* (Oxford: Blackwell, 1997).

24. Putnam, *Bowling Alone*.

25. See, for example, Don Eberly, *America's Promise: Civil Society and the Renewal of American Culture* (Lanham, Md.: Rowman and Littlefield, 1997); Amitai Etzioni, *The Spirit of Community: Rights, Responsibilities and the Communitarian Agenda* (New York: Crown, 1993) and *The New Golden Rule: Community and Morality in a Democratic Society* (New York: Basic Books, 1996).

26. Holloway Sparks, "Dissident Citizenship: Democratic Theory, Political Courage, and Activist Women," *Hypatia: A Journal of Feminist Philosophy* 12 (1997); Iris Marion Young, *Justice and the Politics of Difference* (Princeton, N.J.: Princeton University Press, 1990), and *Inclusion and Democracy* (Oxford: Oxford University Press, 2000).

27. Bickford, *The Dissonance of Democracy*, 10.

28. Manuel Castells, *The City and the Grassroots: A Cross-Cultural Theory of Urban Social Movements* (Berkeley: University of California Press, 1983); Nancy Fraser, "Rethinking the Public Sphere: A Contribution to the Critique of Actually Existing Democracy," in *The Phantom Public Sphere*, ed. Bruce Robbins (Minneapolis: University of Minnesota Press, 1993), 1–32; Young, *Justice and the Politics of Difference*.

29. Monique Deveaux, *Cultural Pluralism and the Dilemmas of Justice* (Ithaca, N.Y.: Cornell University Press, 2000); Miranda Joseph, *Against the Romance of Community* (Minneapolis: University of Minnesota Press, 2002).

30. Little, *The Politics of Community*, 20.

31. Nina Eliasoph, *Avoiding Politics: How Americans Produce Apathy in Everyday Life* (Cambridge: Cambridge University Press, 1998), 244.

32. Quoted in Gregory Alexander, "Dilemmas of Group Autonomy: Residential Associations and Community," *Cornell Law Review* 75 (1989): 13.

33. Carol Greenhouse, Barbara Yngvesson, and David Engel, *Law and Community in Three American Towns* (Ithaca, N.Y.: Cornell University Press, 1994).

34. Bauman, *Community*, 14.

35. Jane Mansbridge, *Beyond Adversary Democracy* (Chicago: University of Chicago Press, 1983). See also Steven Brint, "*Gemeinschaft* Revisited: A Critique and Reconstruction of the Community Concept," *Sociological Theory* 19 (2001): 1–23; and Anne Phillips, "Feminism and the Attractions of the Local," in Rethinking Local Democracy, ed. Desmond King and Gerry Stoker (London: Macmillan, 1996). Both Brint and Phillips note that community groups are often hierarchical and rift-filled, as much as any other organization, and thus are not necessarily effective protectors of equality of voice.

36. Morris Fiorina, "Extreme Voices: A Dark Side of Civic Engagement," in *Civic Engagement in American Democracy*, ed. Theda Skocpol and Morris Fiorina (Washington, D.C.: Brookings Institution Press, 1999), 395–425.

37. Allen Buchanan, "Assessing the Communitarian Critique of Liberalism," *Ethics* 99 (1989): 852–82; Stephen Holmes, "The Permanent Structure of Illiberal Thought," in *Liberalism and the Moral Life*, ed. Nancy Rosenblum (Cambridge: Harvard University Press, 1989), 227–53.

38. Michael Walzer, "The Communitarian Critique of Liberalism," *Political Theory* 18 (1990): 6–23. Will Kymlicka notes that communitarians cannot presume a self incapable of such detachment, because they are themselves involved in a critique of society that requires analytic distancing. Kymlicka, *Contemporary Political Philosophy* (Oxford: Clarendon Press, 1990).

39. As Murray Low notes, this puts communitarians in tension with democratic theory. Even if such theorists profess interest in democratic action, they wish to do so only "on the basis of a consensus of core values which can be specified in advance by academic theorists or public philosophers." Low, "Their Masters' Voice: Communitarianism, Civic Order and Political Representation," *Environment and Planning A* 31 (1999): 87–111, at 106.

40. Fraser, "Rethinking the Public Sphere"; Young, *Justice and the Politics of Difference* and *Inclusion and Democracy*.

41. Barber, *Strong Democracy*, 133.

42. Bickford, *The Dissonance of Democracy*; Bohman, *Public Deliberation*; Chambers, *Reasonable Democracy*; Russell Hanson, "Deliberation, Tolerance and Democracy," in *Reconsidering the Democratic Public*, ed. George Marcus and Russell Hanson (University Park: Pennsylvania State University Press, 1993); Mark E. Warren, *Democracy and Association* (Princeton, N.J.: Princeton University Press, 2001).

43. Fraser, "Rethinking the Public Sphere," 20.

44. Bohman, *Public Deliberation*.
45. Jürgen Habermas, "Three Normative Models of Democracy," *Constellations* 1 (1994): 1–10.
46. Gardbaum, "Law, Politics and the Claims of Community," 713.
47. Castells, *The City and the Grassroots*; Grant McConnell, *Private Power and American Democracy* (New York: Knopf, 1966).
48. Fiorina, "The Dark Side of Civic Engagement."
49. Nancy Rosenblum, *Membership and Morals: The Personal Uses of Pluralism in America* (Princeton, N.J.: Princeton University Press, 1998).
50. Holmes, "The Permanent Structure of Illiberal Thought"; Will Kymlicka, *Liberalism, Community and Culture* (Oxford: Clarendon Press, 1989); J. Donald Moon, *Constructing Community: Moral Pluralism and Tragic Conflicts* (Princeton, N.J.: Princeton University Press, 1993).
51. A strong case for this approach emerges from Paul Hirst, *Associative Democracy: New Forms of Economic and Social Governance* (Amherst: University of Massachusetts Press, 1994).
52. Morris Janowitz, *The Community Press in an Urban Setting: The Social Elements of Urbanism* (Chicago: University of Chicago Press, 1967).
53. Steven Brint even argues that looser groups are more likely to create the virtues of tolerance and equality, because "thicker" communities enforce a conformity that is illegitimately controlling. Brint, *"Gemeinschaft* Revisited."
54. John Ehrenberg, *Civil Society: The Critical History of an Idea* (New York: New York University Press, 1999), 207.
55. John Rawls, *Political Liberalism* (New York: Columbia University Press, 1996).
56. Gerald Doppelt, "Beyond Liberalism and Communitarianism: Towards a Critical Theory of Social Justice," in *Universalism vs. Communitarianism: Contemporary Debates in Ethics,* ed. David Rasmussen (Cambridge, Mass.: MIT Press, 1990), 39–60. The same argument is made by those commentators who emphasize the rise of "neo-liberalism" and its robust emphasis on the market as the central metaphor for society, and its attendant construction of citizens as primarily consumers. See Neil Brenner and Nik Theodore, eds., *Spaces of Neoliberalism: Urban Restructuring in North America and Western Europe* (Oxford: Blackwell, 2002); Wendy Larner, "Neo-Liberalism: Policy, Ideology, Governmentality," *Studies in Political Economy* 63 (2002): 5–25; Jamie Peck and Adam Tickell, "Neoliberalizing Space," *Antipode* 34 (2002): 380–404; Nikolas Rose, *Powers of Freedom: Reforming Political Thought* (Cambridge: Cambridge University Press, 1999).
57. Robert Bellah, Richard Madsen, William Sullivan, Ann Swidler, and Steven Tipton, *Habits of the Heart: Individualism and Commitment in American Life* (New York: Basic Books, 1985); Etzioni, *The Spirit of Community*; Putnam, *Bowling Alone*.

58. As Adrian Little puts it, "Community is sometimes invoked in such an abstract manner that the possibility of practically applying the concept is clearly of secondary importance." Little, *The Politics of Community*, 3.

59. It might be tempting to see this preoccupation with security as a consequence of the political attention given to crime, and the alleged need to wage "war" against it. My analysis of the interviews leads me to conclude that security was understood in broader terms than just protection from crime; hence my emphasis on ontological security. The residents' interest in predictability and reliability appeared to assist them not just to feel physically safe, but also to feel nested in a place of certainty and comfort.

60. Not surprisingly, length of residence is correlated with urban residents' satisfaction with their neighborhood. See Peggy Wireman, *Urban Neighborhoods, Networks and Families* (Lexington, Mass.: Lexington Books, 1984).

61. Robert Sampson verifies that residential stability is a critical factor in establishing connections in neighborhoods, in part to increase the collective capacity to exert informal social control. See Robert Sampson, "Local Friendship Ties and Community Attachment in Mass Society: A Multilevel Systemic Model," *American Sociological Review* 53 (1988): 766–79; "Linking the Micro- and Macro-Level Dimensions of Community Social Organizations," *Social Forces* 70 (1991): 43–64.

CHAPTER TWO

1. Wendy Larner, "Neo-Liberalism: Policy, Ideology, Governmentality," *Studies in Political Economy* 63 (2000): 5–25.

2. Neil Brenner and Nik Theodore, eds., *Spaces of Neoliberalism: Urban Restructuring in North America and Western Europe* (Oxford: Blackwell, 2002); Martin Jones, "Restructuring the Local State: Economic Governance or Social Regulation?" *Political Geography* 17 (1998): 959–88; Alan Lipietz, *Towards a New Economic Order* (Cambridge: Polity Press, 1992); Jamie Peck and Adam Tickell, "Neoliberalizing Space" *Antipode* 34 (2002): 380–404.

3. Rob Atkinson, "Discourses of Partnership and Empowerment in Contemporary British Urban Regeneration," *Urban Studies* 36 (1999): 59–72; David Chandler, "Active Citizens and the Therapeutic State: The Role of Democratic Participation in Local Government Reform," *Policy and Politics* 20 (2001): 219–331; Pat O'Malley, "Risk, Power and Crime Prevention," *Economy and Society* 21 (1992): 252–75; Mike Raco and Rob Imrie, "Governmentality and Rights and Responsibilities in Urban Policy," *Environment and Planning A* 32 (2000): 2187–204; Nikolas Rose, *Powers of Freedom: Reforming Political Thought* (Cambridge: Cambridge University Press, 1999); Kevin Stenson, "Community Policing as a Governmental Technology," *Economy and Society* 22 (1993): 373–89.

4. Richard Briffault, "Our Localism: Part II–Localism and Legal Theory," *Columbia Law Review* 90 (1990): 346–430; Richard Schragger, "The Limits of Localism," *Michigan Law Review* 100 (2001): 371–472. See also Steve Herbert, "The Trapdoor of Community," in *Annals of the Association of American Geographers*, in press.

5. Robert Bellah, Richard Madsen, William Sullivan, Ann Swidler, and Steven Tipton, *Habits of the Heart: Individualism and Commitment in American Life* (New York: Basic Books, 1985); Don Eberly, *America's Promise: Civil Society and the Renewal of American Culture* (Lanham, Md.: Rowman and Littlefield, 1998).

6. Adrian Little, *The Politics of Community* (Edinburgh: Edinburgh University Press, 2002).

7. Suzanne Keller, in her prolonged and intensive examination of a planned neighborhood, discovered that nearly half of the residents she surveyed ranked individual success as very important. By contrast, only nine percent rated community involvement that highly. Keller, *Community: Pursuing the Dream, Living the Reality* (Princeton, N.J.: Princeton University Press, 2003).

8. Carol Greenhouse, Barbara Yngvesson, and David Engel, *Law and Community in Three American Towns* (Ithaca, N.Y.: Cornell University Press); Miranda Joseph, *Against the Romance of Community* (Minneapolis: University of Minnesota Press, 2002); Richard Sennett, *The Uses of Disorder: Personal Identity and City Life* (New York: W. W. Norton, 1972); David Sibley, *Geographies of Exclusion: Society and Difference in the West* (New York: Routledge, 1995); Iris Marion Young, *Justice and the Politics of Difference* (Princeton, N.J.: Princeton University Press, 1990).

9. Mike Davis, *City of Quartz: Excavating the Future in Los Angeles* (New York: Verso, 1990); Setha Low, *Behind the Gates: Life, Security and the Pursuit of Happiness in Fortress America* (New York: Routledge, 2003); Evan McKenzie, *Privatopia: Homeowner Associations and the Rise of Residential Private Government* (New Haven, Conn.: Yale University Press, 1994); Edward Blakely and Mary Snyder, *Fortress America: Gated Communities in the United States* (Washington, D.C. : Brookings Institution Press, 1997).

10. Grant McConnell, *Private Power and American Democracy* (New York: Knopf, 1966).

11. For a more theoretical perspective on the difficulty of cross-cultural political empathy, see Susan Bickford, *The Dissonance of Democracy: Listening, Conflict and Citizenship* (Ithaca, N.Y.: Cornell University Press, 1996). As Bickford notes, there is a perhaps insurmountable challenge involved in trying to communicate politically in a language that is true to one's cultural heritage yet capable of being heard by those in power. As Iris MarionYoung notes, part of the problem here is the restricted means by which political language is accorded legitimacy. Young, *Inclusion and Democracy* (Oxford: Oxford University Press, 2000).

12. The residents in West Seattle are not alone in making these moral assessments of renters. See Keller, *Community*; and Constance Perrin, *Everything in Its Place: Social Order and Land Use in America* (Princeton, N.J.: Princeton University Press, 1977).

13. Robert Sampson shows that residential stability is positively associated with friendship ties and rates of participation in social and leisure activities. Sampson, "Community Attachment in Mass Society: A Multilevel Systemic model," *American Sociological Review* 53 (1988): 766–79, and "Linking the Micro and Macrolevel Dimensions of Community Social Organization," *Social Forces* 70 (1991): 43–64.

14. These resident impressions are supported by evidence that indicates that homeowners are indeed more powerful in localized politics. See Edward Goetz, "Revenge of the Property Owners: Community Development and the Politics of Property," *Journal of Urban Affairs* 16 (1994): 319–34.

15. Indeed, the evidence suggests that relations of trust are most difficult to forge in neighborhoods where disadvantage translates into apprehension about the environment. See Catherine Ross, John Mirowsky, and Shana Pribseth, "Powerlessness and the Amplification of Threat: Neighborhood Disadvantage, Disorder, and Mistrust," *American Sociological Review* 66 (2001): 568–91; Edmund McGarrell, Andrew Giacomazzi, and Quint Thurman, "Neighborhood Disorder, Integration, and the Fear of Crime," *Justice Quarterly* 14 (1997): 479–500.

16. See also Thomas Bender, *Community and Social Change in America* (New Brunswick, N.J.: Rutgers University Press, 1978); Richard Briffault, "Our Localism: Part II"; Matthew Crenson, *Neighborhood Politics* (Cambridge: Harvard University Press, 1983); Stephen Mastrofski, "Community Policing as Reform: A Cautionary Tale," in *Community Policing: Rhetoric or Reality?*, ed. Jack Greene and Stephen Mastrofski (New York: Praeger, 1988), 47–67; Sidney Verba, Kay Schlozman, and Henry Brady, *Voice and Equality: Civic Voluntarism in American Politics* (Cambridge: Harvard University Press, 1995).

17. Morris Fiorina, "Extreme Voices: A Dark Side of Civic Engagement," in *Civic Engagement in American Democracy,* ed. Theda Skocpol and Morris Fiorina (Washington, D.C.: Brookings Institution Press, 1999), 395–425; Young, *Justice and the Politics of Difference.*

18. Verba, Schlozman, and Brady, *Voice and Equality.*

19. Peter Dreier, John Mollenkopf, and Todd Swanstrom, *Place Matters: Metropolitics for the Twenty-first Century* (Lawrence: University Press of Kansas, 2001); Paul Jargowsky, *Poverty and Place: Ghettos, Barrios and the American City* (New York: Russell Sage Foundation, 1997); William Julius Wilson, *The Truly Disadvantaged: The Inner City, the Underclass, and Public Policy* (Chicago: University of Chicago Press, 1990).

20. Manuel Castells, *The City and the Grassroots: A Cross-Cultural Theory of Urban Social Movements* (Berkeley: University of California Press, 1983).

21. J. Donald Moon, *Constructing Community: Moral Pluralism and Tragic Conflicts* (Princeton, N.J.: Princeton University Press, 1993), 203.

22. Nina Eliasoph, *Avoiding Politics: How Americans Produce Apathy in Everyday Life* (Cambridge: Cambridge University Press, 1998).

23. Adam Crawford notes that many crime-prevention efforts are not only narrowly focused, but also relatively short-lived. See Crawford, *The Local Governance of Crime* (Oxford: Clarendon Press, 1997). The short-term nature of such efforts contributed to the lack of enthusiasm toward community policing Randolph Grinc found among the urbanites he interviewed. See Grinc, "'Angels in Marble': Problems in Stimulating Community Involvement in Community Policing," *Crime & Delinquency* 40 (1994): 437–68.

24. Similarly, Theda Skocpol wonders whether local groups can possess much, if any, leverage, especially if their membership lacks wealthy, politically skilled members. Skocpol, "Advocates without Members: The Recent Transformation of American Civic Life," in *Civic Engagement in American Democracy*, ed. Theda Skocpol and Morris Fiorina (Washington, D.C.: Brookings Institution Press, 1999), 461–509.

25. Kay Schlozman, Sidney Verba, and Henry Brady, "Civic Participation and the Equality Problem," in *Civic Engagement in American Democracy*, ed. Theda Skocpol and Morris Fiorina (Washington, D.C.: Brookings Institution Press, 1999), 427–59, at 451.

26. Lawrence Sherman, "Communities and Crime," in *Preventing Crime: What Works, What Doesn't, What's Promising,* ed. Lawrence Sherman (Washington, D.C.: U.S. Department of Justice, 1997), chapter 3.

CHAPTER THREE

1. I obviously oversimplify by referring to "the" state and to "society," as if these were self-contained, unitary entities. In much of this book, I am at pains to demonstrate quite the opposite: that both are complex and diversified. For instance, society is composed of a range of communities, which differ along numerous dimensions, such as location, wealth, education, race, and political capability. Similarly, the state is composed of varied institutions and actors whose relations are rarely unproblematic. However, for the purposes of normative assessment, the broad terms "state" and "society" are useful. They enable me to categorize the types of relations that might pertain between aspects of the state apparatus and members of the citizenry.

2. Iris Marion Young, *Inclusion and Democracy* (Oxford: Oxford University Press, 2000), 5.

3. Max Weber, *Economy and Society: An Outline of Interpretive Sociology*, ed. Guenther Roth and Claus Wittich (New York: Bedminster Press, 1968); Emile Durkheim, *The Division of Labor in Society* (New York: Free Press,

1984); Jürgen Habermas, *Legitimation Crisis* (Boston: Beacon Press, 1975). Peter Manning makes this point eloquently with respect to the police: "The continued deference of citizens to police authority in the absence of specific demands, commands, or laws designed to produce compliance or punish its absence is the source of police power." Manning, "Community Policing as a Drama of Control," in *Community Policing: Rhetoric or Reality?*, ed. Jack Greene and Stephen Mastrofski (New York: Praeger, 1988), 27–45.

4. Not surprisingly, Iris Marion Young highlights her democratic action by recounting her involvement in efforts to establish a civilian review board of police action. Young, *Inclusion and Democracy*.

5. Robert Bursik and Harold Grasmick, *Neighborhoods and Crime: The Dimensions of Effective Community Control* (New York: Lexington Books, 1993); Todd Clear and David Karp, *The Community Justice Ideal: Preventing Crime and Achieving Justice* (Boulder, Colo.: Westview Press, 1999); Lawrence Sherman, "Communities and Crime," in *Preventing Crime: What Works, What Doesn't, What's Promising*, ed. Lawrence Sherman (Washington, D.C.: U.S. Department of Justice, 1997).

6. John Braithwaite, *Crime, Shame and Reintegration* (Cambridge: Cambridge University Press, 1989); John Perry, ed., *Repairing Communities through Restorative Justice* (Lanham, Md.: American Correctional Association, 2002).

7. A comprehensive overview of such efforts, and the issues they raise, can be found in Samuel Walker, *Police Accountability: The Role of Citizen Oversight* (Belmont, Calif.: Wadsworth, 2001).

8. Allen Buchanan, "Assessing the Communitarian Critique of Liberalism," *Ethics* 99 (1989): 852–82; Stephen Gardbaum, "Law, Politics and the Claims of Community," *Michigan Law Review* 90 (1991): 685–760; Will Kymlicka, *Contemporary Political Philosophy* (Oxford: Clarendon Press, 1990); J. Donald Moon, *Constructing Community: Moral Pluralism and Tragic Conflicts* (Princeton, N.J.: Princeton University Press, 1993).

9. John Rawls, *Political Liberalism* (New York: Columbia University Press, 1993).

10. Jeffrey Isaac, Matthew Filner, and Jason Bivins, "American Democracy and the New Christian Right: A Critique of Apolitical Liberalism," in *Democracy's Edges*, ed. Ian Shapiro and Casiano-Hacker Gorden (Cambridge: Cambridge University Press, 1999).

11. Stephen Mastrofski, "Community Policing as Reform: A Cautionary Tale," in *Community Policing: Rhetoric or Reality?*, ed. Jack Greene and Stephen Mastrofski (New York: Praeger, 1988), 47–67.

12. This tension highlights the distinction drawn by Herbert Packer between the due process and crime control models that can ostensibly govern criminal justice practices. In the former, criminal justice officials are strictly regulated by due process considerations and endeavor to follow all procedural rules. In the latter, those officials are encouraged to use their discretion more freely, and to move swiftly and surely against suspected offenders. Police departments that adhere to the due process model risk

losing legitimacy by appearing insufficiently muscular; those more focused on crime control risk arousing concerns about an intrusive state. See Herbert Packer, *The Limits of the Criminal Sanction* (Stanford, Calif.: Stanford University Press, 1968). A more recent version of this debate emerged around the rise of "order maintenance" policing in such American cities as New York and Chicago. For a useful introduction to this debate, see Tracey Meares and Dan Kahan, *Urgent Times: Policing and Rights in Inner-City Communities* (Boston: Beacon Press, 1999).

13. Robert Fogelson, *Big-City Police* (Cambridge: Harvard University Press, 1977); Samuel Walker, *A Critical History of Police Reform* (Lexington, Ky.: Lexington Books, 1977).

14. Peter Dreier, John Mollenkopf, and Todd Swanstrom, *Place Matters: Metropolitics for the Twenty-first Century* (Lawrence: University of Kansas Press, 2001); Rutherford Platt, *Land Use and Society: Geography, Law and Public Policy* (Washington, D.C.: Island Press, 1996); Constance Perrin, *Everything in Its Place: Social Order and Land Use in America* (Princeton, N.J.: Princeton University Press, 1977).

15. Jeffrey Fagan, "Crime, Law and the Community: Dynamics of Incarceration in New York City," in *The Future of Imprisonment,* ed. Michael Tonry (Oxford: Oxford University Press, 2004), 27–59; Becky Petit and Bruce Western, "Mass Imprisonment and the Life Course: Race and Class Inequality in U.S. Incarceration," *American Sociological Review* 69 (2004): 151–69.

16. Michel Foucault, "Governmentality," in *The Foucault Effect: Studies in Governmentality,* ed. Graham Burchell, Colin Gordon, and Peter Miller (Chicago: University of Chicago Press, 1991), 87–104.

17. Mitchell Dean, *Governmentality: Power and Rule in Modern Society* (London: Sage, 1999), 15.

18. Barbara Cruikshank, *The Will to Empower* (Ithaca, N.Y.: Cornell University Press, 1999); Dean, *Governmentality*; Wendy Larner, "Neo-Liberalism: Policy, Ideology, Governmentality," *Studies in Political Economy* 63 (2000): 5–25; Dan MacKinnon, "Managerialism, Governmentality and the State: A Neo-Foucauldian Approach to Local Economic Governance," *Political Geography* 19 (2000): 293–314; Mike Raco and Rob Imrie, "Governmentality and Rights and Responsibilities in Urban Policy," *Environment and Planning A* 32 (2000): 2187–2204; Nikolas Rose, *Powers of Freedom: Reforming Political Thought* (Cambridge: Cambridge University Press, 1999).

19. See also Peter Manning, *Symbolic Communication: Signifying Calls and the Police Response* (Cambridge, Mass.: MIT Press, 1988).

20. James Scott, *Seeing Like a State: How Certain Schemes to Improve the Human Condition Have Failed* (New Haven, Conn.: Yale University Press, 1998).

21. Richard Ericson and Kevin Haggerty document extensively how the various forms police are asked to complete in handling calls structure how officers understand and encode the realities they confront. See their *Policing the Risk Society* (Oxford: Oxford University Press, 1997). Their insightful analysis

overstates the importance of the insurance industry and other external actors in shaping this police epistemology, at least in the context of policing in Seattle and other U.S. cities.

22. Benedict Anderson, *Imagined Communities: Reflections on the Spread of Nationalism* (London: Verso, 1991).

23. Katherine Beckett, *Making Crime Pay: Law and Order in Contemporary American Politics* (Oxford: Oxford University Press, 1997); Stuart Hall, Charles Critcher, Tony Jefferson, John Clarke, and Brian Roberts, *Policing the Crisis: Mugging, the State and Law and Order* (London: Macmillan, 1978).

24. See Steve Herbert, "Morality in Law Enforcement: Chasing 'Bad Guys' with the Los Angeles Police Department," *Law and Society Review* 30 (1996): 799–828.

25. Such a society today is largely unimaginable. As Ian Shapiro states, the "democratic idea is close to nonnegotiable in today's world." Shapiro, *The State of Democratic Theory* (Princeton, N.J.: Princeton University Press, 2003), 1.

26. Malcolm Sparrow, Mark Moore, and David Kennedy, *Beyond 911: A New Era for Policing* (New York: Basic Books, 1990).

27. Herman Goldstein, *Problem-Oriented Policing* (New York: McGraw-Hill, 1991).

28. Will Kymlicka, "Liberal Individualism and Liberal Neutrality," in *Communitarianism and Individualism,* ed. Shlomo Avineri and Avner de-Shalit (New York: Oxford University Press, 1992).

29. The chief's acceptance of the OPA's recommendation was not the only factor in the no-confidence vote. Critical also were his decisions during civil unrest during a Mardi Gras celebration in a downtown region. The chief did not send officers into a melee in the streets, for fear of sparking further turmoil. The decision backfired when one partygoer was bludgeoned to death. The rank and file interpreted the chief's reluctance to send in officers as further evidence that he was too cautious, too fearful of the negative consequences of strong assertions of police authority.

30. Police officers might do well to define politics more expansively, and less defensively. Take, for instance, the definition of politics offered by Stone and his co-authors: "The activity by which a diverse citizenry reconcile, put aside, or in some manner accommodate their differences in order to pursue their common well-being." Clarence Stone, Jeffrey Henig, Bryan Jones, and Carol Pierannunzi, *Building Civic Capacity: The Politics of Reforming Urban Schools* (Lawrence: University of Kansas Press, 2001), 6.

31. Egon Bittner uses more social scientific language to capture the same sentiment: "Police procedure is defined by the feature that it may not be opposed in its course, and that force can be used if it is opposed. This is what the existence of the police makes available to society." Bittner, *The Functions of the Police in Modern Society* (New York: Jason Aronson, 1975), 41.

32. The popularity of so-called "broken-windows" policing, also referred to as "order-maintenance" policing, suggests that perhaps community policing is not as hegemonic as it seems. Although broken-windows policing is often

treated as analogous to community policing, the two are different, largely in terms of the role each envisions for citizen oversight of the police. The popularity of broken-windows policing makes clear that the professional moment is far from over. See Steve Herbert, "Policing the Contemporary City: Fixing Broken Windows or Shoring Up Neo-Liberalism?" *Theoretical Criminology* 5 (2001): 445–66.

33. In an exhaustive review of the literature, Eck and Maguire found little evidence that anything the police do reduces crime. John Eck and Edward Maguire, "Have Changes in Policing Reduced Violent Crime? An Assessment of the Evidence," in *The Crime Drop in America,* ed. Alfred Blumstein and Joel Wallman (Cambridge: Cambridge University Press, 2000), 207–65. The most commonly cited instances of some impact of police actions include so-called "hot spot" policing—where officers concentrate heavy attention on known areas of criminality—and some problem-solving efforts where the police establish an understanding of particular underlying dynamics that can be ameliorated, often through redesign of the physical environment. See Lawrence Sherman, David Farrington, Bandon Welsh, and Denise Gottfredson. eds., *Evidence-Based Crime Prevention* (New York: Routledge, 2002). Whatever successes these strategies might yield, they are notable for the intensiveness of the time and resources they require and for their comparative rarity in the life of a typical urban police department.

34. Peter Manning notes the importance of the police's epistemological construction of the world they inhabit: "The environment in which the officers act is one they largely project, act in accord with, and thus reify." Manning, *Narc's Game: Organizational and Informational Limits on Drug Law Enforcement* (Cambridge, Mass.: MIT Press, 1980), 55.

35. Jewkes and Murcott found, in their analyses of bureaucrats involved in community programs, that these officials typically used metaphors of distance when discussing community. They saw themselves as sharply distinct from the communities with which they worked, and they interpreted their work as acting upon various groups. They also argued that these bureaucrats possessed little sense of what it was like to be inside these communities. Rachel Jewkes and Anne Murcott, "Meanings of Community," *Social Science and Medicine* 43 (1996): 555–63.

36. Manning, *Symbolic Communication.*

37. This scenario can arguably be read as, more than anything, a poorly executed bureaucratic exercise. The dispatcher, for instance, could call the number back to learn more about the situation, and thereby provide more information to the sergeant. Even so, the sergeant's behavior makes plain that he sees this incident in a very particular way, one conditioned by his experience as a police officer. Indeed, the strength of his orientation evidently makes it difficult for him to accept a decision by the wife that she may view as a simple and understandable mistake.

CHAPTER FOUR

1. A good example is the resistance officers mounted to "team policing," an
    attempt by many American police departments in the 1970s to introduce
    many of the principles—decentralization of police authority, collective
    problem solving—later included in community policing. See Jack Greene,
    "Police and Community Relations: Where Have We Been and Where Are We
    Going?" in *Critical Issues in Policing*, ed. Roger Dunham and Geoffrey Alpert
    (Prospect Heights, Ill.: Waveland Press, 1989), 349–68; Mark Harrison Moore,
    "Problem-Solving and Community Policing," in *Modern Policing*, ed. Norval
    Morris and Michael Tonry (Chicago: University of Chicago Press, 1992).
2. This is not to suggest that all officers rebuff community policing. There is
    support for it within the Seattle Police Department and in other police
    departments. Eugene Paoline, for instance, found that the strongest
    differentiating attitude between the police officers he surveyed was their
    orientation toward community policing. Paoline, "Shedding Light on
    Police Culture: An Examination of Officers' Occupational Attitudes," *Police
    Quarterly* 7 (2004): 205–36. Yet resistance remains strong, such that little
    change seems evident as a consequence of its alleged institutionalization.
    See William Lyons, *The Politics of Community Policing: Rearranging the Power
    to Punish* (Ann Arbor: University of Michigan Press, 1999); Edward
    Maguire, "Structural Change in Large Municipal Police Organizations
    during the Community Policing Era," *Justice Quarterly* 14 (1997): 701–30.
    This resistance deserves explanation, in no small part because it has
    enormous consequences for the prospect of improved police-community
    relations.
3. Egon Bittner, *The Functions of the Police in Modern Society* (New York: Jason
    Aronson, 1975).
4. It is a mistake, of course, to posit that there is a single, unified entity called
    "police culture." As with any system of meaning, police culture is
    differentiated. See Steve Herbert, "Police Subculture Reconsidered,"
    Criminology 36 (1998): 334–70. Further, as with all social actors, police
    officers vary in the extent to which their actions are shaped by elements of
    that culture. For instance, officers take different stances toward community
    policing; some embrace it, while others are hostile. See Eugene Paoline,
    Stephanie Myers, and Robert Worden, "Police Culture, Individualism, and
    Community Policing: Evidence from Two Police Departments," *Justice
    Quarterly* 17 (2000): 575–605; and Janet Chan, *Changing Police Culture:
    Policing in a Multicultural Society* (Cambridge: Cambridge University Press,
    1997). That said, hostility to community policing is quite common, and is
    palpable in Seattle. Indeed, this hostility leads some commentators to
    argue that police culture is the biggest impediment to the wider
    implementation of the community policing philosophy. See Malcolm
    Sparrow, Mark Moore, and David Kennedy, *Beyond 911: A New Era for*

Policing (New York: Basic Books, 1990); Susan Sadd and Randolph Grinc, "Innovative Neighborhood Oriented Policing: An Evaluation of Community Policing Programs in Eight Cities," in *The Challenges of Community Policing: Testing the Promises,* ed. Dennis Rosenbaum (Thousand Oaks, Calif.: Sage Publications, 1994), 27–52.

5. I provide a more complete account of police culture elsewhere: Herbert, "Police Subculture Reconsidered." My goal here is to isolate elements of that culture that are particularly relevant to the resistance to community policing.

6. Anthony Bouza, *The Police Mystique* (New York: Plenum Press, 1990).

7. In Los Angeles, officers make the contrast between bold and cowardly officers by employing two terms—"hard charger" and "station queen." The former's willingness to embrace danger is valued, the latter's desire to lay low is ridiculed. See Steve Herbert, "Hard Charger or Station Queen? Policing and the Masculinist State," *Gender, Place and Culture* 8 (2001): 55–71.

8. Ibid.

9. Sadd and Grinc note that the community police officers they interviewed felt compelled to argue continually with their colleagues that they were engaged in "real" police work. Sadd and Grinc, "Innovative Neighborhood Oriented Policing."

10. A pat-down is legally distinguishable from a full search. The former is acceptable as long as the officer can demonstrate a plausible concern about safety. The latter can only occur if the officer possesses either probable cause that a crime has occurred or a court-sanctioned search warrant.

11. David Bayley suggests that this process begins when the officer suits up for work. As he puts it, "Police officers especially in the United States go to work as if they were going to war." Bayley, *Police for the Future* (New York: Oxford University Press, 1994), 70.

12. Oberweiss and Musheno, in a study contrasting police officers and vocational rehabilitation counselors, noticed an important difference in the way the two groups justified the discretion they valued. Counselors defended discretion because it enabled the flexibility they thought they needed to generate positive outcomes for their clients. By contrast, the police legitimated discretion because it preserved their freedom to exercise their authority as they wished. Trish Oberweiss and Michael Musheno, *Knowing Rights: State Actors' Stories of Power, Identity, and Morality* (Aldershot: Ashgate, 2001).

13. The lack of supervision here was especially surprising because the suspended license policy was the subject of a public controversy initiated primarily by members of minority groups. These activists argued that the policy disproportionately punished poor people and worked to exacerbate their economic challenges by making transportation more difficult. The

officer thus contributed, albeit probably unwittingly, to an erosion of police legitimacy in the eyes of those who suspect officers of racial profiling. Because of this controversy, a greater supervisory hand might have been expected.

14. The police's propensity to explain crime in largely individualist terms is noted, as well, in Adam Crawford, *The Local Governance of Crime* (Oxford: Clarendon Press, 1997), and Stuart Scheingold, *The Politics of Street Crime: Criminal Process and Cultural Obsession* (Philadelphia: Temple University Press, 1991). Scheingold notes that this individualist approach leads officers to favor deterrence as their preferred punishment approach and to eschew more societal explanations for crime.

15. The broken windows logic holds that visible signs of "disorder" send a signal to would-be criminals that a place suffers from inadequate informal social control and thus represents an opportunity for misdeeds. See James Q. Wilson and George Kelling, "Broken Windows," *Atlantic Monthly,* March 1982, 29–38; George Kelling and Catherine Coles, *Fixing Broken Windows: Restoring Order and Reducing Crime in Our Cities* (New York: Free Press, 1996). Although the title of the theory emphasizes the built environment, in actuality its focus is upon people. As Wilson and Kelling put it in their seminal essay, they are concerned specifically with "disreputable or obstreperous or unpredictable people: panhandlers, drunks, addicts, rowdy teenagers, prostitutes, loiterers, the mentally disturbed." Not surprisingly, they endorse a strong police response to such street denizens, including close surveillance and frequent arrests. The theory has spawned a movement within policing, referred to as either broken windows or order-maintenance, that has been used to legitimate stringent crackdowns on activities associated with the "disreputable" people that concern them, most notably in the "zero tolerance" strategy advocated in New York City in the 1990s. For a defense of this by the police chief who supervised it, see William Bratton, *Turnaround: How America's Top Cop Reversed the Crime Epidemic* (New York: Random House, 1998). For critiques, see Bernard Harcourt, *The Illusion of Order: The False Promise of Broken Windows Policing* (Cambridge: Harvard University Press, 2001); Steve Herbert, "Policing the Contemporary City: Fixing Broken Windows or Shoring up Neo-Liberalism?" *Theoretical Criminology* 5 (2001): 445–66.

16. Recall the discussion in chapter 3 of the community police officer in Blufftop who sought authority to make arrests for trespassing. He sought this authority because he suspected that the loiterers were delivering drugs. To arrest them for drug delivery, however, would require a high level of probable cause. It would be much simpler for him to arrest them for trespassing.

17. It seems to have escaped the notice of the typical patrol officer that the United States now has the highest rate of incarceration in the world.

18. Ralph Saunders, "The Space Community Policing Makes and the Body that Makes It," *The Professional Geographer* 51 (1999): 135–46.

CHAPTER FIVE

1. This is not uncommon. Residents of largely African-American neighborhoods often complain not just about overly aggressive policing, but about an inability to compel a police response when they desire it. See Randall Kennedy, *Race, Crime and the Law* (New York: Pantheon, 1997).

2. In their comprehensive and imaginative exploration of public sentiments, Loader and Mulcahy note the deep well of support for the police in the United Kingdom. As they put it, "As an institution intimately concerned with the viability of the state and the security of its citizens . . . policing remains closely tied to people's sense of ontological security and collective identity, and capable of generating high, emotionally-charged levels of identification among citizens." Ian Loader and Aogan Mulcahy, *Policing and the Condition of England* (Oxford: Oxford University Press, 2003).

3. This reflects a commonly reported difference between assessments of the police by whites and by blacks; the latter see the police in consistently more negative terms. See David Bayley and Harold Mendelsohn, *Minorities and the Police* (New York: Free Press, 1968); John Hagan and Celesta Albonetti, "Race, Class and the Perception of Criminal Injustice in America," *American Journal of Sociology* 88 (1982): 329–55; Herbert Jacob, "Black and White Perceptions of Justice in the City," *Law and Society Review* 6 (1971): 69–90; Ronald Weitzer, "Citizens' Perceptions of Police Misconduct: Race and Neighborhood Context," *Justice Quarterly* 16 (1999): 819–46.

4. James Gilsinan describes this process this way: "Information is not simply gathered. It is worked with and transformed so that heterogeneous, ambiguous, and sometimes emotionally overlaid requests for service can be simplified and made orderly." As he notes, the 911 operators he studied necessarily slotted citizen requests into certain categories to make them comprehensible and actionable. Gilsinan, "They Is Clowning Tough: 911 and the Social Construction of Reality," *Criminology* 27 (1989): 329–44.

5. See the discussion of class and community power in chapter 2.

CHAPTER SIX

1. Wesley Skogan and Jeffrey Roth, "Introduction," in *Community Policing: Can It Work?*, ed. Wesley Skogan (Belmont, Calif.: Wadsworth/Thomson Publishing, 2004), xvii–xxxiv. One could counter-argue that policing labeled under the rubric of either "broken windows" or "order maintenance" is equally significant, particularly given its alleged—and heavily politicized—relationship to the reduction of crime in such U.S. cities as New York. Even so, community policing is obviously of monumental significance to modern policing.

2. See Jack Greene, "Community Policing and Organizational Change," in *Community Policing: Can It Work?*, ed. Wesley Skogan (Belmont, Calif.: Wadsworth/Thomson Publishing, 2004), 30–53; Steve Herbert, "Hard Charger or Station Queen? Policing and the Masculinist State," *Gender, Place and Culture* 8 (2001): 55–71; Richard Wood, Mariah Davis, and Amelia Rouse, "Diving into Quicksand: Program Implementation and Police Subcultures," in *Community Policing: Can It Work?* (Belmont, Calif.: Wadsworth/Thomson Publishing, 2004), 136–61. As Greene (2004, 48) summarizes the situation, "It is clear that the police imagination remains captured by nineteenth-century ideas about crime and police response." To be sure, this line of critique about the culture of resistance needs to be balanced against assertions that community policing possesses some popularity within police organizations, and that it possesses potential to effect further change in police culture. See Eugene Paoline, Stephanie Myers, and Robert Worden, "Police Culture, Individualism, and Community Policing: Evidence from Two Police Departments," *Justice Quarterly* 17 (2000): 575–605; Janet Chan, *Changing Police Culture: Policing in a Multicultural Society* (Cambridge: Cambridge University Press, 1997); Robert Kane, "Permanent Beat Assignment in Association with Community Policing: Assessing the Impact of Officers' Field Activity," *Justice Quarterly* 17 (2000): 259–80. That said, it is a safe assumption that support for community outreach efforts has always existed, to some extent, within the ranks of officers. Similarly, as my analysis in Seattle suggests, moves to implement community policing may work less to change police culture than to *intensify the resistance* to change.

3. John Crank, "Watchman and Community: Myth and Institutionalization in Policing," *Law & Society Review* 28 (1994): 325–51; Matthew Giblin, "Institutional Theory and the Recent Adoption and Activities of Crime Analysis Units in U.S. Law Enforcement Agencies," Ph.D. dissertation, Department of Criminal Justice, Indiana University, 2004; Jihong Zhao, *Why Police Organizations Change: A Study of Community-Oriented Policing* (Washington, D.C.: Police Executive Research Forum, 1996). A comprehensive survey of American police departments did show, however, that those departments that professed to be engaged in community policing were more likely to have features associated with it. See Edward Maguire and Charles Katz, "Community Policing, Loose Coupling, and Sense Making in American Police Agencies," *Justice Quarterly* 19 (2002): 503–36.

4. Edward Maguire, "Structural Change in Large Municipal Police Organizations during the Community Policing Era," *Justice Quarterly* 14 (1997): 547–76; Dennis Rosenbaum and Deanna Wilkinson, "Can Police Adapt? Tracking the Effects of Organizational Change over Six Years," in *Community Policing: Can It Work?*, ed. Wesley Skogan (Belmont, Calif.: Wadsworth/Thomson Publishing, 2004), 79–108.

5. William Lyons, *The Politics of Community Policing: Rearranging the Power to Punish* (Ann Arbor: University of Michigan Press, 1999); Wesley Skogan and Susan Hartnett, *Community Policing, Chicago Style* (New York: Oxford University Press, 1997).

6. See Skogan and Hartnett, *Community Policing, Chicago Style.*

7. Skogan and Hartnett note that the vast majority—90 percent—of proposed solutions to problems raised at community-police forums in Chicago are generated by the police. Ibid.

8. Wesley Skogan, Susan Hartnett, Jill DuBois, Jennifer Comey, Karla Twedt-Ball, and Erik Gudell, *Public Involvement: Community Policing in Chicago* (Washington, D.C.: National Institute of Justice, 2000), 5.

9. Jeffrey Roth, *National Evaluation of the COPS Program: Title I of the 1994 Crime Act* (Washington, D.C.: National Institute of Justice, U.S. Department of Justice, 2000), 237. This pessimistic assessment of improved police-community relations extends more broadly to community crime prevention programs, which show no appreciable impact on crime or community empowerment. See Tim Hope, "Community Crime Prevention," in *Building a Safer Society: Strategic Approaches to Crime Prevention,* ed. Michael Tonry and David Farrington (Chicago: University of Chicago Press, 1995).

10. William Lyons notes that community policing was never initiated anywhere as a citizen-sponsored effort, but always at the behest of the police. It is therefore perhaps not surprising that police departments resist significant citizen interaction. Lyons, *The Politics of Community Policing.*

11. To be sure, skepticism attended much early assessment of the prospects of community policing. See, most notably, Jack Greene and Stephen Mastrofski, eds., *Community Policing: Rhetoric or Reality?* (New York: Praeger, 1988).

12. Ralph Taylor, *Breaking Away from Broken Windows: Baltimore Neighborhoods and the Nationwide Fight against Crime* (Boulder, Colo.: Westview Press, 2001).

13. Eric Klinenberg, "Bowling Alone, Policing Together," *Social Justice* 28 (2001): 75–80.

14. Ibid., 80. It is also critical to note the longstanding tensions between the police and many minority-dominated neighborhoods. This legacy of distrust is not easily overcome. It also breeds a strong desire within these neighborhoods for robust forms of citizen oversight of the police, something officers resist. See Randolph Grinc, "Angels in Marble: Problems in Stimulating community involvement in community policing," *Crime & Delinquency* 40 (1994): 437–68; Lyons, *Politics of Community Policing;* Skogan and Hartnett, *Community Policing, Chicago Style.*

15. Peter Manning makes this point evocatively: "As long as the police exercise authority and violence in the name of the state, they will be feared and loathed by some segments of the community and will represent this violence potential to all segments of the community." Manning, "Community Policing as a Drama of Control," in *Community Policing:*

Rhetoric or Reality?, ed. Jack Greene and Stephen Mastrofski (New York: Praeger, 1988), 34.

16. Adam Crawford, *The Local Governance of Crime* (Oxford: Clarendon Press, 1997).

17. See Anthony Pate, "Experimenting with Foot Patrol: The Newark Experience," in *Community Crime Prevention: Does It Work?*, ed. Dennis Rosenbaum (Beverly Hills, Calif.: Sage, 1986), 137–>56; Wesley Skogan, *Disorder and Decline: Crime and the Spiral of Decay in American Neighborhoods* (Berkeley and Los Angeles: University of California Press, 1990). Certainly, residents in West Seattle agreed that they would prefer more visible patrol, an indication that such visibility would increase their sense of security.

18. It seems plausible to blame police administrators for the failure to institutionalize community policing reforms. Perhaps these administrators failed to change the reward structure enough to give officers incentives to reform, or perhaps they failed to communicate the underlying philosophy in sufficient detail. See Susan Sadd and Randolph Grinc, "Innovative Neighborhood Oriented Policing: An Evaluation of Community Policing Programs in Eight Cities," in *The Challenges of Community Policing: Testing the Promises,* ed. Dennis Rosenbaum (Thousand Oaks, Calif.: Sage Publications, 1994), 27–52; Skogan and Hartnett, *Community Policing, Chicago Style*. There is undoubted merit to these arguments, yet the example of Chief Stamper in Seattle suggests that these administrative efforts, however well managed, will face deep and powerful resistance. The reality and power of this resistance suggests that the continued hope for reform through community policing is misplaced.

19. See Steve Herbert, "Policing the Contemporary City: Fixing Broken Windows or Shoring up Neo-liberalism?" *Theoretical Criminology* 5 (2001): 445–66; Peter Kraska and Victor Kappeler, "Militarizing American Police: The Rise and Normalization of Para-Military Units," *Social Problems* 44 (1997): 1–18.

20. Katherine Beckett, *Making Crime Pay: Law and Order in Contemporary American Politics* (New York: Oxford University Press, 1997).

21. See Rebecca Petit and Bruce Western, "Mass Imprisonment and the Life Course: Race and Class Inequality in U.S. Incarceration," *American Sociological Review* 69 (2004): 151–69.

22. Jeffrey Fagan, "Crime, Law and Community: Dynamics of Incarceration in New York City," in *The Future of Imprisonment,* ed. Michael Tonry (Oxford and New York: Oxford University Press, 2004), 27–60.

23. For an assessment of problem-solving policing, see John Eck, "Why Don't Problems Get Solved?" in *Community Policing: Can It Work?*, ed. Wesley Skogan (Belmont, Calif.: Wadsworth/Thomson Publishing, 2004), 185–206.

24. George Berkley, *The Democratic Policeman* (Boston: Beacon Press, 1969).

25. Ian Loader and Aogan Mulcahy, *Policing and the Condition of England* (Oxford: Oxford University Press, 2003).

26. Robert Sampson and Janet Lauritsen, "Violent Victimization and Offending: Individual, Situational and Community-Level Risk Factors," in *Understanding and Preventing Violence,* ed. Albert Reiss and Jeffrey Roth (Washington, D.C.: National Academy of Sciences, 1993). See also Alice O'Connor, "Swimming against the Tide: A Brief History of Federal Policy in Poor Communities," in *Urban Problems and Community Development,* ed. Ronald Ferguson and William Dickens (Washington, D.C.: Brookings Institution Press, 1999), 77–137.

27. Peter Dreier, John Mollenkopf, and Todd Swanstrom, *Place Matters: Metropolitics for the Twenty-first Century* (Lawrence: University of Kansas Press, 2001). Clarissa Hayward makes a similar argument in the case of education. Her ethnography of power in classroom settings showed her that the realities teachers faced were determined by conditions created by state policies concerning zoning, taxation, and housing. In other words, their teaching strategies depended upon the demographic characteristics of their classrooms, a reality created by state policies that enable segregation. Not surprisingly, the ways teachers in advantaged neighborhoods approached their students differed from those used by teachers in disadvantaged areas. Clarissa Rile Hayward, *De-facing Power* (Cambridge: Cambridge University Press, 2000).

28. Lisa Miller argues that the citizens she interviewed in her assessment of Weed and Seed operations in central Seattle did not want police at the center of community building. She quotes one community leader: "We didn't need the police to save our community. We could do it ourselves." Lisa Miller, *The Politics of Community Crime Prevention* (Aldershot: Ashgate, 2001), 176.

29. This is not to suggest that the police's tactics are irrelevant to the lived realities of urbanites, particularly in distressed neighborhoods. Law enforcement tactics are geographically concentrated. For instance, in Seattle, enforcement of drug trafficking laws is focused on particular areas, namely those reputed for sales of crack cocaine. These practices generate arrest rates for African Americans that are grossly disproportionate to their presence in the Seattle population. See Katherine Beckett, "Race, Drug Delivery, Law Enforcement: Lessons from Seattle," unpublished manuscript. As noted above, these geographically variant enforcement practices produce staggering impacts in a period of massive incarceration in the United States. Neighborhoods where arrest rates are high lose large numbers of young males to prison, with a cascade of negative consequences. See Fagan, "Crime, Law and Community"; Petit and Western, "Mass Imprisonment and the Life Course."

30. Matthew Crenson and Benjamin Ginsberg, *Downsizing Democracy: How America Sidelined Its Citizens and Privatized Its Public* (Baltimore: Johns Hopkins University Press, 2002). The state is also critical in creating the political opportunity structure citizen organizations face and seek to manipulate. This structure critically shapes what organizations can and cannot do, and thus the strategies they can expect to see succeed or fail. For

analysis at the national level, see Stanley Tarrow, *Power in Movement: Social Movements, Collective Action and Politics* (New York: Cambridge University Press, 1994); for analyses at the local level see Byron Miller, *Geography and Social Movements: Comparing Antinuclear Activism in the Boston Area* (Minneapolis: University of Minnesota Press, 2000); Randy Stoecker and Anna Vakil, "States, Cultures and Community Organizing: Two Tales of Two Neighborhoods," *Journal of Urban Affairs* 22 (2000): 439–58.

31. John Ehrenberg, *Civil Society: The Critical History of an Idea* (New York: New York University Press, 1999), 231.

32. Robert Bellah and his associates rightly note that part of the issue here is the status accorded the state. Too much suspicion of the state deflates civil action before it starts. This suspicion, as I have demonstrated in the case of community policing, is not without merit. Yet its persistence can be self-defeating. As they cogently put it: "The transformation of the state, however complex that process would be, should focus on bringing a sense of citizenship into the operation of government itself. Such a spirit is not entirely lacking today, but it is severely weakened by suspicion of government and politics on the one hand and the idea of impersonal efficient administration on the other. In order to limit the danger of administrative despotism, we need to increase the prestige of the state, not derogate it." Robert Bellah, Richard Madsen, William Sullivan, Ann Swidler, and Steven Tipton, *Habits of the Heart: Individualism and Commitment in American Life* (New York: Basic Books, 1985), 211.

33. Allen Buchanan uses this point to make the case that communitarians needlessly downplay the importance of the rights liberalism defends. He argues that the protection of individual rights provides the space in which communities can flourish. Thus, he suggests, liberalism can be justified on community as much or more than on individual grounds. If the impulse to community is so strong, then it will flourish in a rights-protected environment. Buchanan, "Assessing the Communitarian Critique of Liberalism," *Ethics* 99 (1989): 852–82.

34. Take, for example, state policies done in the name of reducing terrorism. Couched in the language of minimizing the threat of the "evil" of terror—itself a term that connotes irrational malevolent action—state policies can be legitimated even if they run afoul of such traditional concerns as the protection of civil liberties.

35. This tension between economic competitiveness and communal togetherness is recognized by a number of scholars. See Bellah et al., *Habits of the Heart*; Andre Gorz, *Capitalism, Socialism, Ecology* (London: Verso, 1994); Adrian Little, *The Politics of Community: Theory and Practice* (Edinburgh: Edinburgh University Press, 2002). The construction of individuals as consumers also lies in some tension with their existence as rights-bearing individuals. See Gerald Doppelt, "Beyond Liberalism and Communitarianism: Towards a Critical Theory of Social Justice," in

Universalism vs. Communitarianism: Contemporary Debates in Ethics, ed. David Rasmussen (Cambridge, Mass.: MIT Press, 1990), 39–60. On the specific contradictions between communitarianism and neo-liberalism, see Steve Herbert, "The Trapdoor of Community," *Annals of the Association of American Geographers,* forthcoming.

36. This reality helps explain why so many commentators place such faith in community as a means to fill the needs the state leaves unmet. See Pierre Clavel, Jessica Pitt, and Jordan Yin, "The Community Option in Urban Policy," *Urban Affairs Review* 32 (1998): 435–58.

37. Nina Eliasoph, *Avoiding Politics: How American Produce Apathy in Everyday Life* (Cambridge: Cambridge University Press, 1998), 250.

38. Renditions of successful stories can be found in Peter Medoff and Holly Sklar, *Streets of Hope: The Fall and Rise of an Urban Neighborhood* (Boston: South End Press, 1994); Mark R. Warren, *Dry Bones Rattling: Community Building to Revitalize Democracy* (Princeton, N.J.: Princeton University Press, 2001).

39. See Peter Dreier, "Community Empowerment Strategies: The Limits and Potential of Community Organizing in Urban Neighborhoods," *Cityscape: A Journal of Policy Development and Research* 2 (1996): 121–59; O'Connor, "Swimming against the Tide"; William Sites, *Remaking New York: Primitive Globalization and the Politics of Urban Community* (Minneapolis: University of Minnesota Press, 2003).

40. Morris Fiorina, "Extreme Voices: A Dark Side of Civic Engagement," in *Civic Engagement in American Democracy,* ed. Theda Skocpol and Morris Fiorina (Washington, D.C.: Brookings Institution Press, 1999), 395–425; Jane Mansbridge, *Beyond Adversary Democracy* (Chicago: University of Chicago Press, 1983).

41. Richard Briffault, "Our Localism: Part II–Localism and Legal Theory," *Columbia Law Review* 90 (1990): 346–430; Dreier, Mollenkopf, and Swanstrom, *Place Matters.*

42. Gary Delgado, *Organizing the Movement: The Roots and Growth of ACORN* (Philadelphia: Temple University Press, 1986); Dreier, "Community Empowerment Strategies"; Margaret Weir, "Power, Money and Politics in Community Development," in *Urban Problems and Community Development,* ed. Ronald Ferguson and William Dickens (Washington, D.C.: Brookings Institution Press, 1999), 139–92.

43. Robert Halperin's history of neighborhood-level efforts to reduce the consequences of poverty reveals just this danger. As he puts it: "The history of neighborhood initiative reflects American society's persistent tendency to ask those who have the least role in making and the largest role in bearing the brunt of society's economic and social choices to deal with the effects of those choices." Halperin, *Rebuilding the Inner City: A History of Neighborhood Initiatives to Address Poverty in the United States* (New York: Columbia University Press, 1995), 5.

44. Theda Skocpol, "Advocates without Members: The Recent Transformation of American Civic Life," in *Civic Engagement in American Democracy,* ed. Theda Skocpol and Morris Fiorina (Washington, D.C.: Brookings Institution Press, 1999), 461–509.

45. Robert Putnam, *Bowling Alone: The Collapse and Revival of American Community* (New York: Simon and Schuster, 2000).

46. The importance for local groups of expanding their political struggles to other scales is shown consistently in the literature that analyzes community organizing. See Manuel Castells, *The City and the Grassroots* (Berkeley: University of California Press, 1983); Robert Chaskin, Prudence Brown, Sudhir Venkatesh, and Avis Vidal, *Building Community Capacity* (New York: Aldine de Gruyter, 2001); Dreier, "Community Empowerment Strategies"; Ronald Ferguson, "Social Science Research, Urban Problems, and Community Development Alliances," in *Urban Problems and Community Development,* ed. Ronald Ferguson and William Dickens (Washington, D.C.: Brookings Institution Press, 1999), 569–610; Robert Fisher, *Let the People Decide: Neighborhood Organizing in America* (New York: Twayne Publishers, 1994); Stites, *Remaking New York;* Warren, *Dry Bones Rattling.*

47. Peggy Wireman makes this point forcefully: "Informal support systems do not have the power or resources to handle many of the basic problems of individuals, families and neighborhoods, which are caused by national, and even international, economic, political, and social forces. It is naive, if not callous, to expect informal systems to alleviate or solve problems caused by the adverse effects of these forces." Peggy Wireman, *Urban Neighborhoods, Networks and Families* (Lexington, Mass.: Lexington Books, 1984),146. It is also naive to expect poor neighborhoods to organize themselves to ward off the economic calamities that beset them. See Richard Cloward and Frances Fox Piven, "Disruptive Dissensus: People and Power in the Industrial Age," in *Reflections on Community Organization: Enduring Themes and Critical Issues,* ed. Jack Rothman (Itasca, Ill.: F. E. Peacock, 1999).

48. Even to the extent that place-based politics do generate any capacity to help improve the circumstances facing poor neighborhoods, they always compete, ideologically, with discourses of social betterment that de-emphasize place. This is particularly true in the neo-liberal era, where the economic market and its allegedly ineluctable and universal logic is said to confer benefits across the social landscape, regardless of pre-existing conditions. See O'Connor, "Swimming against the Tide." Further, many place-based political programs in contemporary cities are focused on economic development rather than social welfare. This often increases political tensions between local groups, as some favor this development while others resist it. See Margit Meyer, "Urban Movements and Urban Theory in the Late 20th Century," in *The Urban Moment,* ed. Sophie Body-Gendrot and Bob Beauregard (Thousand Oaks, Calif.: Sage, 1999), 209–39. At the neighborhood level, property owners often dominate in ways that do

not necessarily, or even likely, converge with the agendas of those who seek to support the disadvantaged. See Susan Fainstein and Clifford Hirst, "Urban Social Movements," in *Theories of Urban Politics,* ed. David Judge, Gerry Stoker, and Harold Wolman (Thousand Oaks, Calif.: Sage Publications, 1995), 181–203.

49. See Warren, *Dry Bones Rattling.*

50. See Robert Sampson, Stephen Raudenbush, and Felton Earls, "Neighborhoods and Violent Crime: A Multilevel Study of Collective Efficacy," *Science* 277 (August 1997): 918–24.

51. Clarence Stone, Jeffrey Henig, Bryan Jones, and Carol Pierannunzi, *Building Civic Capacity: The Politics of Reforming Urban Schools* (Lawrence: University of Kansas Press, 2001).

Index